MW01623678

# EXPLORING CREATION WITH
# ASTRONOMY
## 2nd EDITION

*The heavens are telling of the glory of God; And their expanse is declaring the work of His hands. Psalm 19:1*

**Author**
**Jeannie K. Fulbright**

**Technical Editor and Content Contributor**
**Damian R. Ludwiczak**

**Exploring Creation with Astronomy**
**$2^{nd}$ Edition**

Published by
Apologia Educational Ministries, Inc.
P.O. Box 896844
Charlotte, NC, 28289-6844
www.apologia.com



ISBN: 978-1-940110-58-5

Cover design: Doug Powell
Book design: Andrea Kiser Martin
Cover photos courtesy NASA/ JPL/ Caltech

All Biblical quotations are from
the New American Standard Bible (NASB)
unless otherwise noted.

Printed by Asia Printing Co., Ltd, Seoul, South Korea
March 2021

10 9 8 7 6

## Apologia's Young Explorer Series
# INSTRUCTIONAL SUPPORT

Apologia's elementary science materials launch young minds on an educational journey to explore God's signature in all of creation. Our award-winning curriculum cultivates a love of learning, nurtures a spirit of exploration, and turns textbook lessons into real-life adventures.

### TEXTBOOK

**Apologia Textbooks** are written directly to the student in a highly readable conversational tone. Periodically asking students to stop and retell what they have just heard or read, our elementary science courses engage students as active learners while growing their ability to communicate clearly and effectively. With plenty of hands-on activities, the Young Explorer Series allows young scientists to actively participate in the scientific method.

### NOTEBOOKING JOURNAL

Spiral-bound **Apologia Notebooking Journals** contain lesson plans, review questions, full-color mini-books, puzzles, and much more to keep students actively engaged in learning while keeping them organized.

### JUNIOR NOTEBOOKING JOURNAL

Designed for younger students and those who struggle with writing, **Apologia Junior Notebooking Journals** cover everything in the regular notebooking journals, but at a more basic level and with primary writing lines. With simpler vocabulary pages, and additional coloring pages, junior notebooking journals make science enjoyable for even your youngest student.

### AUDIO BOOKS

Some students learn best when they can see and hear what they are studying. Having the full audio text of your course is great for listening while reading along in the book or riding in the car! **Apologia Audio Books** contain the complete text of the book read aloud to your student.

At Apologia, we believe in homeschooling. We are here to support your endeavors and to help you and your student thrive! Find out more at apologia.com.

# TABLE OF CONTENTS

## Lesson 3:



## Lesson 4:



## Lesson 5:



## Lesson 6:

## Lesson 7:



## Lesson 8:



## Lesson 9:



## Lesson 10:

**Lesson 11:**



**Lesson 12:**



**Lesson 13:**



**Lesson 14:**

# INTRODUCTION TO
# ASTRONOMY

## wisdom from above

*"It is I who made the earth, and created man upon it. I stretched out the heavens with My hands And I ordained all their host."*

Isaiah 45:12

## Textbook

Thank you for choosing *Exploring Creation with Astronomy, 2nd edition* as your science course this year. This textbook was designed to launch your mind on an educational journey through God's incredible universe. It will bring the cosmos into your home through the exploration of the major structures of our solar system, including the sun, planets, and asteroid belt. It then journeys into the stars and galaxies to explain how marvelous and mighty our Creator is.

## Activities

The *Exploring Creation with Astronomy, 2nd edition* course is carefully designed to provide fun, educational, and safe activities for you to conduct in your home. We have tested these activities in multiple home settings. Accidents, however, can occur anywhere and at any time. Therefore, we urge you to follow safe experimental processes at all times. Be sure adult supervision is provided for all activities. Follow standard safety procedures by ensuring use of eye protection and gloves if needed. Set aside equipment to be used only in experiments. Never experiment and cook or eat with the same items, especially if you are using inedible or potentially toxic materials. You should NEVER eat any item in an activity unless the lesson instructions are for an edible activity AND your parents have read the lesson, checked the materials, and given permission.

## Course Website

When you look at the night sky, what you will see depends on where you are located on Earth. As a result, it is difficult to write an astronomy book that applies to everyone. In addition, it would be hard to continually update a book to give you the future dates for interesting events. To get the most out of this course, you should regularly visit the course website. It is designed to link you to multimedia that relates to astronomy. Please always practice safe Internet use. While we monitor our book extras, we do link to outside sources, and we cannot guarantee each site on a day-to-day basis.

Please go to apologia.com and click on the Book Extras link to find instructions for accessing our Book Extras material. For this course, you will need the following password:

**Godcreatedtheheavens**

## Notebooking Journals

The 14 lessons in this book are in-depth and contain quite a bit of scientific information and hands-on activities. Each lesson should be broken up into manageable time slots depending on a child's age and attention span; this will vary from family to family. A suggested lesson plan is located in your notebooking journal. Whether you are using the regular or junior notebooking journal, this journal will be your place to document your studies in astronomy, record results of your activities, and create colorful memories.

# Summary

When you read science textbooks you are studying important information discovered by great scientists throughout history. Keeping a student notebooking journal helps to document studies in your own words. Additionally, participating in science activities builds upon that knowledge as you conduct your own scientific studies. It is our prayer that at the end of the year you will have gained a deeper appreciation for and understanding of the beauty of creation.

*Teach me Your way, O LORD;*
*I will walk in Your truth;*
**Psalm 86:11**

LESSON 1

# WHAT IS ASTRONOMY?

## *wisdom from above*

There is order to our universe, and once you start to understand that order, science will not be a class, but rather a means to recognize your Creator's signature throughout creation.

*For since the creation of the world His invisible attributes, His eternal power and divine nature, have been clearly seen, being understood through what has been made, so that they are without excuse.*

Romans 1:20

# Welcome

When you look up at the night sky, what do you see? Everything that you can see around you and up in the sky, and even things that you *can't* see, are part of our universe. The universe is everything that exists! That includes all of the planets (even Earth), stars, galaxies, and intergalactic space.

***Isn't God as high as the heavens?***
***And look at the highest stars—how lofty they are!***
**Job 22:12**

The study of outer space is called **astronomy** (uh strahn' uh me). The word *aster* means star, while *onomy* means knowledge of. The word *astronomy*, then, means knowledge of the stars. Many years ago, the only word used for every object in outer space was *aster* or star. In other words, every light in the night sky was called a star. We still use the word *astronomy* to talk about the study of everything in space, even though the way we use it today means more than just studying the stars.

***There is one glory of the sun, and another glory of the moon, and another glory of the stars;***
***for star differs from star in glory.* 1 Corinthians 15:41**

An **astronomer** is someone whose job is to study the stars, the planets, and everything else in outer space. You are going to be a novice (beginner) astronomer this year because you will be studying the universe as you take this course.

## The Night Sky

Have you ever been out in the countryside at night far away from city lights? On a clear night in the countryside, you can see many thousands of stars in the sky. It's truly a miraculous sight, and it is called the *Milky Way*. It's the galaxy we live in. At the center of the picture to the right, 2 bright objects are visible. The brightest is the planet Jupiter, while the other is the star Antares. The red laser beam points to the center of our galaxy. You will learn more about our galaxy later in this course.

The night sky on a clear night away from city lights.

## *think about this*

The Bible tells us that God made the stars and Moon to give us light at night and a calendar to follow. He also uses the night sky to give us signs to mark important events. Scientists have also learned that the planets, stars, and many other things in space help to keep life going on Earth.

*"Then God said, 'Let there be lights in the expanse of the heavens to separate the day from the night, and let them be for signs and for seasons and for days and years; and let them be for lights in the expanse of the heavens to give light on the earth'; and it was so."* Genesis 1:14–15

It's easy to confuse planets and stars.

## Stars and Planets

Not everything you see in the night sky that shines like a star actually is a star. Some of the brightest objects in the night sky (besides the Moon), for example, are planets. Stars appear to twinkle in the sky, but planets do not twinkle. There are also stars that appear to move rapidly across the sky and then disappear. We call them shooting stars, but they are not stars at all. They are meteors (mee' tee orz). We will talk more about all of these later.

**Tell someone in your own words what the universe is. Can you remember what the word *astronomy* means?**

# Clocks, Calendars, and Seasons

Did you know that the sun, Moon, planets, and stars in the sky help us tell time, create our calendar with days and years, and help us know the timing of the seasons? It's true! Many years ago, before people had clocks and calendars, they told the time of day by the position of a shadow on the ground. They also knew when a month had passed by looking at the shape of the Moon in the night sky.

This sundial uses a shadow to tell time.

Phases of the Moon.

Stonehenge

There is an ancient stone structure in southern England called Stonehenge. Many believe that ancient people used it to tell when spring had arrived. They judged the season by the position of the sun in relation to the large stones that make up the structure. Knowing when the seasons arrive helped them to time the planting and harvesting of crops.

6th century mosaic

The man-named patterns of stars in the night sky are called **constellations** (kahn' stuh lay' shuns). Ancient people knew which constellations would be in the sky in each season of the year—winter, spring, summer, or fall. They also used the constellations to mark what year it was and how many years had passed since an event. The picture here is very old. It shows the constellations around the central sun. The corners show the 4 turning points of the year. Many years ago, then, before we had calendars in our homes, the night sky marked the passage of time.

**Take a moment to tell someone in your own words what you have learned so far. You can use the illustrations to help you remember and to show them examples of what you are talking about.**

## think about this

A miraculous sign given to man was when God placed a star in the sky over the city of Bethlehem indicating the Savior had come. When wise men from a distant land saw the star, they used the star to navigate their travels to Bethlehem to see and worship Jesus. *"They went their way; and the star, which they had seen in the east, went on before them until it came and stood over the place where the Child was. When they saw the star, they rejoiced exceedingly with great joy. After coming into the house they saw the Child with Mary His mother; and they fell to the ground and worshiped Him. Then, opening their treasures, they presented to Him gifts of gold, frankincense, and myrrh."* Matthew 2: 9-11

## Navigation

A long time ago, the sun, planets, and stars also helped sailors know which direction to sail. This was called **celestial** (suh les' chul) **navigation**. Today sailors use compasses and Global Positioning Systems (GPS). A compass is a device with a needle that always points to the magnetic north. GPS uses satellites in outer space to track your position on Earth. You always know what direction you are going if you have a compass; you know exactly where you are if you have GPS. If you have neither of these, you can still know which direction you are going if you know the positions of the stars!

God's plan for the lights in the sky does not only include mankind.

Astrolabes helped ancient sailors navigate by using the position of the stars.

Scientists have learned that some birds know to fly south for the winter by the constellations. This is called migration. God made a very special way for birds to know when and how to fly south. He created within them a special gift we call **instinct** (in' stinkt). One instinct that God has given birds tells them to look at the constellations to know when to migrate south for the winter and when to migrate back north for the summer. It also tells them how to use the constellations to know which direction they must fly. This is why some birds often fly at night when they migrate.

**Can you explain in your own words about navigating with stars, a compass, and GPS? Tell someone about the gift of instinct and how some birds use the stars to navigate. Who else used a star to navigate to a very special event?**

## Gravity

Our solar system is made up of the sun, 8 planets and their moons, dwarf planets, asteroids, comets, and meteoroids. Some planets are actually very important to our home planet, Earth. Although their effect is mathematically small, these planets help to fasten Earth in place. They keep Earth from moving too far away from the sun or too close to it. In other words, the planets keep our world steady. You see, the sun pulls on Earth with a force called gravity.

An artist's drawing of our solar system.

**Gravity** is an invisible force that pulls objects toward each other. When we drop something, it doesn't really fall; it gets pulled down to Earth by gravity. Instead of saying that "it fell," it would be more scientifically correct to say, "It was pulled to Earth." All the planets and their moons have gravity. Larger planets have more gravity than smaller planets. The sun, the largest object in our whole solar system, has the most gravity of all. God placed the planets and sent them to circle (orbit) around the sun at the perfect distance.

If the planet Mercury (mur' kyur ree) were very much closer to the sun, or if the sun were very much larger, Mercury would get pulled into the sun. Instead, it stays exactly where God put it because it has been placed at the right distance from the sun. The pull that planets (and other objects) have on each other is called **gravitational pull**. The sun, Earth, and all of the planets have gravitational pull. Earth's gravitational pull on the Moon keeps the Moon where it is. The Moon's gravitational pull on Earth makes the oceans bulge as it passes by. The sun's gravitational pull keeps the planets in their places in the solar system.

### *think about this*

Isaac Newton was a very important and influential scientist. He once said, "Gravity explains the motions of the planets, but it cannot explain who sets the planets in motion." We must never forget that when we are studying science, we are learning about things of which God already knows because He created them!

# Our Solar System

This is a drawing that represents our solar system. Only part of the sun is shown, and each planet is shown along with where it is in relation to the sun. As you can see, Mercury is closest to the sun, while Neptune is farthest from the sun. The relative sizes of the planets are fairly accurate; however, the distance between the planets is not correct.

The 8 planets in our solar system are Mercury, Venus (vee' nus), Earth, Mars, Jupiter, Saturn, Uranus (yur' uh nuhs), and Neptune (nep' toon). This is also the order in which they travel around the sun.

A fun way that many people remember the planets and their order is by using a mnemonic (nih mahn' ik). The first letter of each planet is made into a different word that makes a sentence. Look at this example:

| **M**ercury | **V**enus | **E**arth | **M**ars | **J**upiter | **S**aturn | **U**ranus | **N**eptune |
|---|---|---|---|---|---|---|---|
| **M**y | **V**ery | **E**ducated | **M**other | **J**ust | **S**erved | **U**s | **N**achos |

Notice that the word underneath each planet begins with the first letter of the planet's name. "My very educated mother just served us nachos" is a silly sentence, but the first letter of each word in that sentence helps you remember the order of the planets.

# Activity 1.1

## Create Your Own Mnemonic

| **M**ercury | **V**enus | **E**arth | **M**ars | **J**upiter | **S**aturn | **U**ranus | **N**eptune |
|---|---|---|---|---|---|---|---|
| | | | | | | | |

For each box below a planet, choose a word that begins with the first letter of that planet. Try to make a sentence that you will remember. It will be easier to remember a sentence that makes sense. You can also have fun creating silly mnemonics. Place them in your notebooking journal.

**Can you explain in your own words what you have learned about gravity and the solar system?**

# Astronomers and Astronauts

An astronaut in a space suit.

There were many people in history who have helped us understand astronomy better. In the year 1510, a man named **Nicolaus** (nik' oh lus) **Copernicus** (koh pur' nih kus) had the unusual and amazing idea that Earth revolved around the sun. At that time, everyone thought that all the stars and planets revolved around Earth. We now know that Copernicus was correct, even though most people during his time did not believe him.

**Galileo** (gal ih lay' oh) **Galilei** (gal ih lay') was an astronomer who believed Copernicus. He taught how to use telescopes to study the planets and stars, and many of the observations that he made helped scientists understand that Copernicus was right about the sun being at the center of our solar system. Galileo was able to learn a lot of things about our solar system through the wonderful telescopes he built.

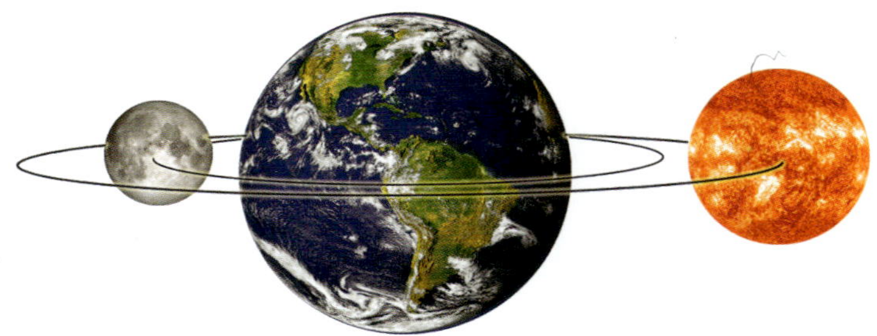

In Copernicus's time, everyone thought that Earth was the center of the solar system and that all of the planets and the sun revolved around it.

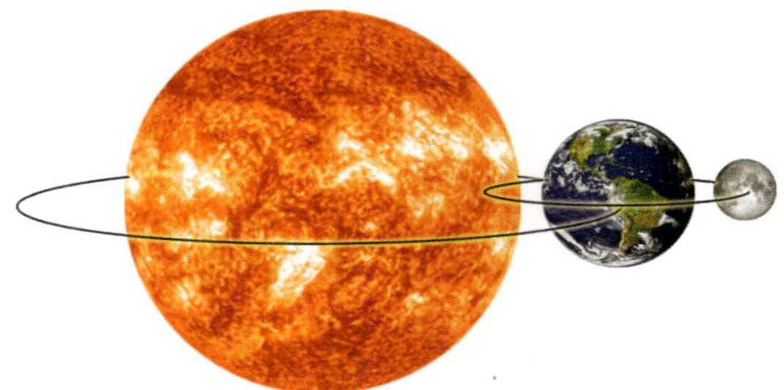

Copernicus thought that a more elegant arrangement of the solar system would be for the sun to be at the center and for the planets to revolve around the sun.

Today, a lot of astronomers work for **NASA**. NASA is America's space agency, and it stands for **N**ational **A**eronautics and **S**pace **A**dministration. If you want to be an astronomer when you grow up, you might want to work for NASA. It is also the organization that sends people and spaceships to space. If you like to build and invent things, you could be a NASA engineer. The picture on the next page shows a rocket being built by NASA engineers. See how tiny the engineers at the bottom of the picture are? That gives you an idea of how big the rocket is. Many NASA engineers build spaceships, telescopes, robots, and other useful things for space exploration. As you go through this book, you will learn about NASA spacecraft used to explore the solar system and the universe.

If you become an astronaut, you will probably work for NASA. An **astronaut** is someone who is trained to travel in a spaceship into outer space. Astronauts wear special spacesuits to explore outer space. Maybe one day you will be an astronaut and go to some of the places we will study in this course!

This is a picture of the *Saturn V* rocket being assembled by NASA engineers. The *V* in *Saturn V* is the Roman numeral 5, which refers to the number of engines in the rocket's first stage.

The Hubble Space Telescope in orbit around Earth.

Have you ever looked through a telescope? You can see a long way off when you do. You will see many pictures that come from telescopes as you study this course. There is an enormous telescope floating up in space that sends pictures back down here to Earth. It is called the Hubble Space Telescope. Even though a telescope will make a planet look like it is much closer, most planets can be seen without a telescope if you know where to look.

**Use your own words to tell someone what you know about astronomers and astronauts. Also tell them about NASA and what it does.**

NASA has more than a dozen Earth science satellites in orbit. These satellites help NASA scientists study the oceans, land, and atmosphere.

## Satellites

Something else you can see in the night sky are satellites. A **satellite** is an object in space that travels in circles around another object. The Moon is a satellite of Earth because it travels in a circle around the planet. So when you are looking up in the sky, you can say, "Oh look! I see a satellite!" as you point at the Moon.

An **artificial satellite** is made by man and sent into space to orbit around Earth. *Artificial* means not natural or something that is made by human hands. Only God makes **natural satellites**. People have sent thousands of artificial satellites to travel around Earth. These satellites do many jobs. Some can look closely at any part of the world and take pictures for others to see. Some can put a lot more channels on your TV. Some can look at planets and stars far away. Some watch the weather and take pictures so the weatherman can tell us if it is going to rain. Artificial satellites are very important. If you look up into the sky at night and see a small point of light (like a star) that is moving across the sky, you are probably looking at a satellite. You can go to **www.apologia.com/bookextras** for links to several websites to discover where satellites are above Earth at this very moment. One may even be over your house!

**Can you tell someone what you have learned about satellites?**
**Are you able to explain the difference between a natural and an artificial satellite?**

# Activity 1.2
## Build a Model Solar System

Scientists use models to show a concept. The model of the solar system you are going to build will not be to scale, which means that the size of the sun and planets, as well as their distance from each other, will not be exact.

### You will need:

- Adult supervision
- Balloons of many sizes and colors
- Construction Paper
- Markers
- Measuring Tape
- Thread, ribbon, or string
- Scissors
- Thumbtacks or tape

### You will do:

1. You will start by choosing several balloons, each of which will represent a planet. For each planet, try to choose a balloon with a color that is something like the planet's color. Mercury should be somewhat gray, for example, while Earth should be blue. Use the pictures in this book to help you decide the color for each planet.

**(Continued on next page.)**

2. Once you have chosen a balloon for a planet, slowly blow up the balloon. As you blow it up, measure the distance across the diameter of the balloon at its widest point. Tie the balloon closed when the diameter is close to the number on the chart.

| Planet the Balloon Represents | Diameter of the Balloon |
|---|---|
| Mercury | 1 inch |
| Venus | 2 1/2 inches |
| Earth | 2 5/8 inches |
| Mars | 1 3/8 inches |
| Jupiter | 29 1/5 inches |
| Saturn | 25 inches |
| Uranus | 10 5/8 inches |
| Neptune | 10 1/4 inches |

**NOTE:**
If you decide to include the sun, have your student measure 300 inches in order to recognize that the sun's diameter is far larger than any of the planets' diameters.

3. If you do not have a long, thin balloon for Saturn's ring, make a circle out of construction paper that will fit around the balloon that represents Saturn.
4. Label your planets.
5. Tie a string, ribbon, or thread to each balloon, and hang them from the ceiling using thumbtacks or tape. Make sure you hang the balloons in the correct order.

**Your solar system model is now complete. Great job!**

## What Do You Remember?

Why did God create the stars and planets? What are the names of the planets? Do you remember the name of the astronomer who first said that Earth revolves around the sun? What is the name of the astronomer who learned how to study space with a telescope? What is the name of America's space agency and what does it do? What is the difference between a natural and an artificial satellite? What was your favorite part of this lesson?

LESSON 2
# THE SUN

## wisdom from above

As we begin our wonderful journey through our solar system, it's important to keep in mind that our God created it all to demonstrate His glory and everlasting love.

*To Him who made the great lights,*
*For His loving kindness is everlasting:*
*The sun to rule by day,*
*For His loving kindness is everlasting,*
*The moon and stars to rule by night,*
*For His loving kindness is everlasting.*
*Give thanks to the God of heaven,*
*For His loving kindness is everlasting.*

Psalm 136:7-9, 26

# The Star of Stars

The sun seems so small up in the sky, but it is one of the biggest things God created. It is so enormous, so gigantic, that you have never seen anything so big in all of your life. It's bigger than the biggest thing on Earth. What's the biggest thing you can think of that you have seen on Earth? Did you know that the sun is millions of times bigger than that? The sun makes everything on Earth, and even some things in space, look like tiny little specks of dust or tiny little ants. The difference in size between Earth and the sun is very great. You can see the difference if you compare a small peppercorn to a basketball. If you don't have a basketball, a dinner plate will work also. Just remember that the sun is not flat like a plate.

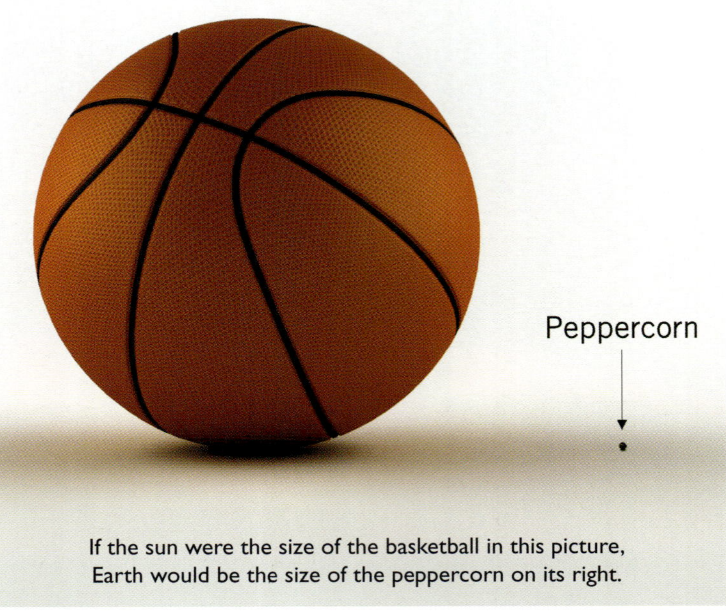

If the sun were the size of the basketball in this picture, Earth would be the size of the peppercorn on its right.

Now look at these 2 objects next to each other. The peppercorn represents Earth, and the basketball or plate represents the sun. Amazing isn't it? You would need more than one million Earths to fill up the sun! The whole Earth is like a tiny speck compared to the sun. If Earth is so small compared to the sun, imagine a mountain on Earth compared to the sun. Imagine a person compared to the sun. Imagine an ant compared to the sun!

Why does the sun seem small in the sky if it is so much bigger than Earth? Shouldn't it look big to us since we are tiny in comparison?

| Solar System Planet | How Many Planets to Fill Up the Sun |
|---|---|
| Mercury | 23,171,635 |
| Venus | 1,518,766 |
| Earth | 1,297,126 |
| Mars | 8,582,021 |

| Solar System Planet | How Many Planets to Fill Up the Sun |
|---|---|
| Jupiter | 921 |
| Saturn | 1,537 |
| Uranus | 20,153 |
| Neptune | 22,157 |

This table describes how many planets it takes to fill up the sun.

# Activity 2.1
## Understanding Distance and Size

### You will need:
- Your eyes
- One finger

### You will do:
1. Look out a window if you are inside. If you are outside look as far away as you can.
2. Find something on Earth, such as a house or tree or mountain that is really far away, as far away as you can see.
3. Close one of your eyes and hold your finger up next to that far-away object.
4. Doesn't it look like that object is smaller than your finger?

### Discussion
You know that the object is much larger than your finger, but your finger seems bigger is because it's closer. The closer an object is to us, the bigger it will seem. Even big objects look tiny when they are far away.

## 92,956,050
Can you say that number? It will help if you call the first comma million and the second comma thousand. That is a giant number. *Ninety-two million, nine hundred fifty-six thousand fifty* is how we say it. That's how many miles we are away from the sun. That is further than you can even imagine. Most people round up and just say 93,000,000 miles.

How many miles did you travel last time you went on a trip? Did you go 100 miles? 700 miles? I'll bet you didn't go 93,000,000 miles. To give you an idea of how far that is, if you could drive to the sun as fast as you can on the highway, it would take you about 163 years to get there! That would be a very long trip. But we wouldn't want to take a trip to the sun. Do you know why? It's because the sun is just too hot!

The sun is a light with heat and swirling gases. God made the sun so hot that we can feel its heat millions of miles away. The hottest your oven can get is about 500 °F (260 °C). The sun is usually about 10,000 °F (5538 °C) on the surface and 27,000,000 °F on the inside! If something as hot as the sun touched Earth, it would burn a hole all the way through in an instant. God put the sun the perfect distance from Earth. If it were closer, the water in the oceans would evaporate. The trees and plants would all die, and we would all burn up. If the sun were farther away, its warming energy would not reach us, so the oceans would freeze into big blocks of ice, and we would too.

**Can you explain in your own words what you have learned so far?
Why does the sun look so small if it is really so very big?**

## Don't Stare!

The sun's light is so powerful that it isn't safe to look at the sun! **Staring at the sun (or any extremely bright object) can cause terrible damage to your eyes.** Have you ever looked at a bright light bulb and then had to look away after a moment? The sun is about 4 septillion times brighter than a light bulb. A septillion has 24 zeros! This is why you can injure your eyes if you look directly at the sun. To study the sun, scientists look at it with the help of special tools.

Did you know that you can take a magnifying glass out into the sunlight and concentrate its energy? This works especially well during the summer when the sun is shining its light more directly upon us. When you focus the sunlight coming through the curved lens of the magnifying glass for a short period of time, the rays of light are bent by the curved lens and concentrated to a small spot, focusing all of the heat and energy there. That's a lot of heat and energy to be going to one spot. Guess what! You have a lens just like a magnifying glass in your eye. If you look at the sun, your eye-lens will concentrate the sun's light and focus it to a very small spot on the back of your retina (the back wall of your eye.) This can cause permanent eye damage or blindness. Also, there are no pain sensors in your retina, so you wouldn't even know it is happening. **Never look directly at the sun!**

# Activity 2.2

## Use a Magnifying Glass to Focus Heat

### You will need:

- Adult supervision
- Magnifying glass
- A 1-inch pad of butter (or substitute chocolate)
- Paper plate
- Sunny day

### You will do:

1. Cut a stick of butter so that you have a piece about 1-inch thick on your paper plate.
2. Take your plate, butter, and your magnifying glass outside.
3. Set the plate on the ground, and hold the magnifying glass in your hand.
4. Point the magnifying glass down at the ground away from the butter. Make sure that light from the sun is hitting the magnifying glass.

(Continued on next page.)

5. Move the magnifying glass up and down, and play with how you are holding it, until you see a circle of light on the ground. As the rays from the sun travel through the magnifying glass lens, they are concentrated into a circle. Depending on how close the magnifying glass is to the ground, you should see the spot of light change size. Practice adjusting the size of the spot by moving the magnifying glass up and down.
6. Use the magnifying glass to make a spot of light hit the butter. Try to make the spot of light as small as possible. Watch what happens for the next few minutes.
7. Do the same thing, but this time, make the spot of light a little bigger. Once again, watch what happens.

### Discussion

What did you see in the experiment? You should have noticed that the butter didn't melt much when it just sat out in the sun. However, as soon as you used the magnifying glass, the butter should have melted quickly. Why? The magnifying glass concentrated the energy of the sun's light into a small spot. That heated up the butter, which made it melt. You should also have noticed that the smaller the spot, the faster the butter melted. That's because the smaller the spot, the more concentrated the energy. That resulted in even more heat, which melted the butter more quickly.

## *think about this*

When Moses asked to see God, God said, *"You cannot see My face, for no man can see Me and live"* Exodus 33:20, So Moses hid behind some rocks when God passed by. Even though he just got a glimpse of God's glory, Moses's face glowed for a long time. He had to wear a veil over his face so that the Israelites would not be afraid of him. The Bible describes angels and the transfigured Jesus as being very bright, like a light. The sun is impressive, but we know God is even more magnificent. The sun gives us a small clue to the splendor of God.

# Activity 2.3

## Give a Speech to Teach Others About the Sun

Write (or dictate) a speech about why people should not look at the sun. You may want to begin with some interesting facts about the sun that you have learned so far. Be sure to explain what happens to your retina when you look at the sun. You can also make illustrations. Practice your speech several times. Try to memorize most of it so you will not sound like you are reading when you give your speech to others. It often helps to write one important key word from each sentence on an index card. Then you just glance at the key word and remember what your next sentence is. This will help you to sound like you are *talking* to your audience and not *reading* to your audience. Once you have practiced the speech, give your speech to help teach others what you have learned.

## Revolve and Rotate

When one object travels in a circle around another object, it is orbiting that object. In other words, it is revolving around that object. We use the words *orbit* and *revolve* to mean the same thing. One object is in the center, while the other object moves in a circle around it.

Many years ago, astronomers and everyone else believed that the sun revolved around Earth. Every morning they would see the sun coming up in the east and going down in the west. It just seemed logical that it was circling around Earth.

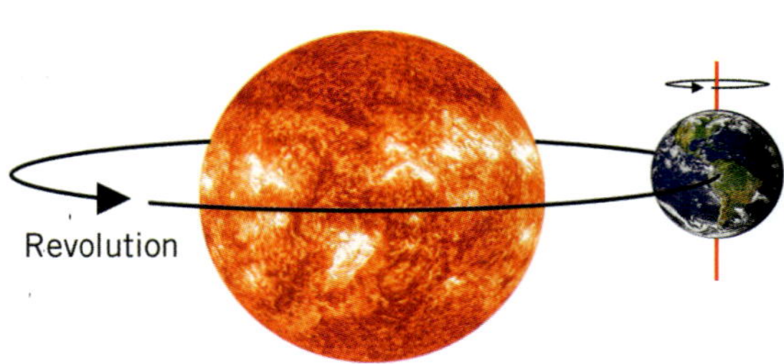


Earth *revolves* around the sun as it *rotates*.

A drawing of our solar system. The planets orbit around the sun in paths that are almost (but not quite) circles. While they orbit around the sun, they also rotate. A planet's rotation causes night to turn into day.

But now we know that the sun is in the center of the solar system, and Earth revolves, or orbits, in a circle around the sun. Remember that a satellite is an object in space that revolves around another object. Earth is a natural satellite of the sun, just as the Moon is a natural satellite of Earth! The sun has many natural satellites, including the 8 planets, several dwarf planets, and thousands of asteroids, comets, and meteoroids. Earth has only one natural satellite (the Moon), but there are thousands of artificial satellites orbiting Earth.

# Activity 2.4a

## Take a Walk Around the Sun - Revolving

### You will need:

- 3 people

### You will do:

1. Make 1 person the sun.
2. Have someone be Earth and walk in a circle (revolve) around the sun.
3. Now you be the Moon and walk in a circle (revolve) around Earth while Earth circles the sun. The same side of the Moon always faces Earth, so while you are the Moon, be certain that you are always facing the person who is Earth.
4. Do this several times until you can do it easily. The planets and their moons are doing this every day, all day long!

Every time a planet goes all the way around the sun and comes back to the same spot, it has completed one revolution around the sun. We would say it has done one orbit or has revolved one time. When Earth completes one revolution around the sun, we say that a year has passed.

## *think about this*

Have you ever heard it said that someone's life or world revolves around something or someone? One person might say, "Her life revolves around ballet." Another person might say, "That mother's world revolves around her baby." This expression means that something or someone is very important. What does your life revolve around? It is best for all our lives to revolve around Jesus, for He should be the Ruler of our lives!

Did you know that as the planets revolve around the sun, they are also spinning around at the same time? This is called rotating. They are rotating as they orbit the sun.

# Activity 2.4b

## Take a Walk Around the Sun - Rotating

Now try this little adventure: Just as before, put someone in the middle of the floor to be the sun. Now you pretend that you are a planet. Before you begin to walk around the sun, start turning around and around (rotating) in place. Just twirl around in place, and then begin to walk around the sun while you twirl around at the same time. See if you can get the third person to be the Moon and revolve around you as you both rotate and revolve around the sun! Go slow so you don't get too dizzy or bump into each other.

As Earth *rotates*, night turns into day.

Did you notice that while you were twirling around, sometimes you faced the sun and sometimes you faced away from the sun? It's the same with the planets! One part is facing the sun; and as the planet rotates, this same part is facing away from the sun. When we face the sun, it's very bright and light. We call this *day*. When we are facing away from the sun, there isn't any light shining on us. We call this *night*. One side of the planet is always facing the sun, so it is always day somewhere on Earth. Every planet rotates, so every planet has a day and a night. Even the Moon has a day and a night.

This spinning around, or rotating, is happening right now! We are rotating while we revolve around the sun, and God made it so that we don't even get dizzy! Try not to confuse the terms *revolve* and *rotate*. Rotating is the spinning that makes it day and night. You can think of it this way: "It rotates between day and night." Revolving is the orbiting around the sun that takes a year. Remember that people say someone's life revolves around things. It's easy to remember that someone's life would revolve around that thing for a whole year.

**Can you explain in your own words the difference between rotating and revolving?**

# Solar Flares, Auroras, and Sunspots

The sun is very active. The fire on the sun is jumping and hopping and rolling about, something like a campfire does. It is a huge ball of fire that is spinning around and around. In the large picture of the sun on the left, you can see fire darting out from the sun. That is called a **solar flare**. It is a giant tower of fire that is many times larger than any planet in our solar system. Look at the picture on the left. See how small Earth is compared to the solar flare? That gives you some idea of how big the flare is! Solar flares burst out millions of miles from the sun. The solar flares throw so much energy and electricity toward Earth that people in the far north and the far south can see colorful electrical lights up in the sky at night caused by these flares. These lights are called **auroras** (uh roar' uhs). Sometimes a solar flare can make a burst of energy come through a city's power lines so forcefully that the entire city can lose its power and electricity. Solar flares can also cause problems in telephones and radios.

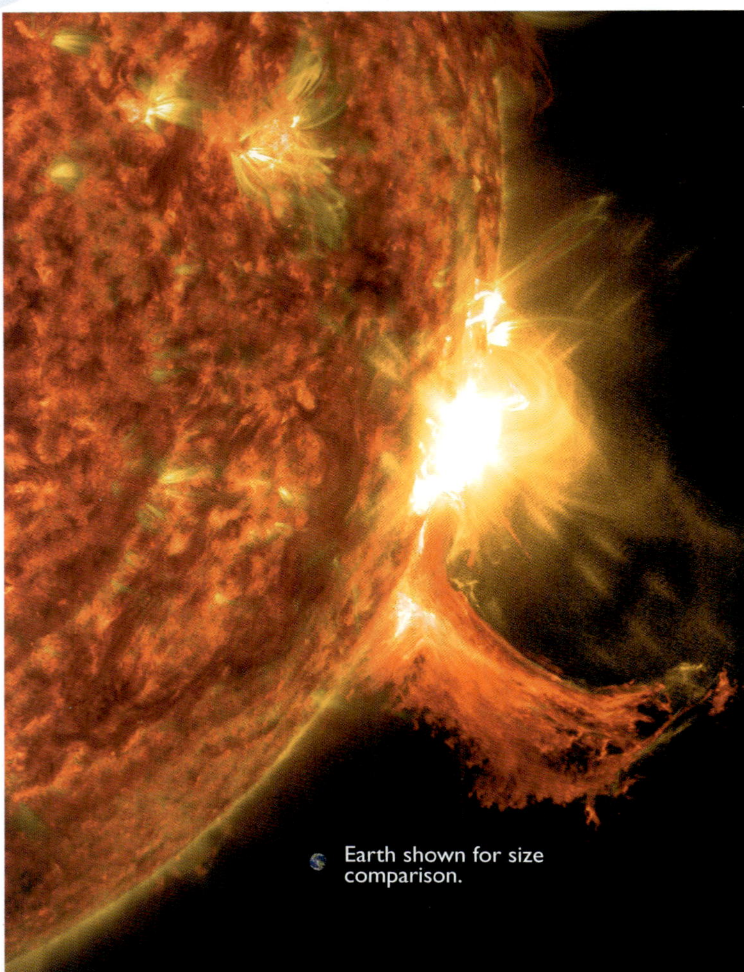

A photograph of a solar flare coming off the sun. An image of Earth has been placed in the picture to give you an idea of how big the flare is.

Look at the picture of the sun below. It was taken with special tools used to study the sun. Do you notice anything interesting about this picture? There are little dark spots on the sun. They are called **sunspots**. They are spots that are cooler than the rest of the sun. They are about 7,000 °F. That's still very hot, but much cooler than the rest of the outer part of the sun. Do you see the dark spot pointed out in the photograph? That spot is bigger than Earth! Some are so big that they are 10 times bigger than Earth.

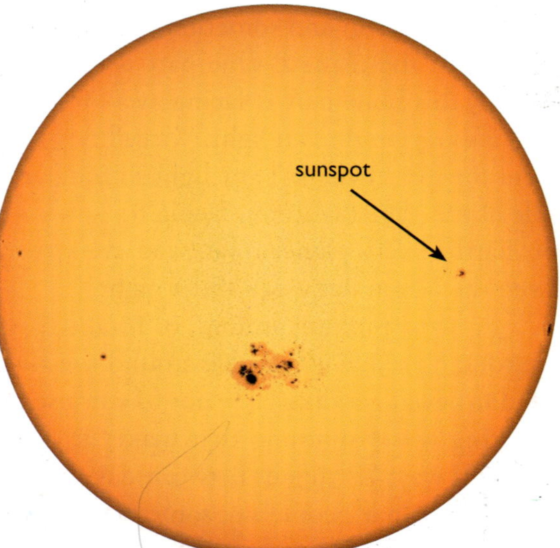

A photograph of the sun. The dark spots are sunspots.

Many scientists believe that sunspots affect the weather here on Earth. You see, when the number of sunspots on the sun is large, the sun is actually *hotter* than when the number of sunspots is small. This may seem strange to you since the sunspots themselves are cooler than the rest of the sun. However, scientists think that sunspots result in more activity in the sun, so the sun actually gets hotter when there are more sunspots. Without sunspots Earth would be cooler than it is today. This happened a few hundred years ago (1645–1710). During those years, scientists could not see any sunspots on the sun, and Earth was many degrees cooler than normal. In the same way, if there were too many sunspots, Earth would get really hot, and rain would not fall as often. This would cause severe droughts (drouts) on Earth. Droughts are long seasons with no rain. When it doesn't rain, the plants die. Without the plants, many animals die as well. People also depend on rain to live. The food we eat cannot grow without rain. A drought can be a very dangerous thing for life on Earth. God has designed our sun with sunspots to help Earth stay at the right temperature. Everything in the universe is God's design. He has taken special care with every feature of the universe to protect us.

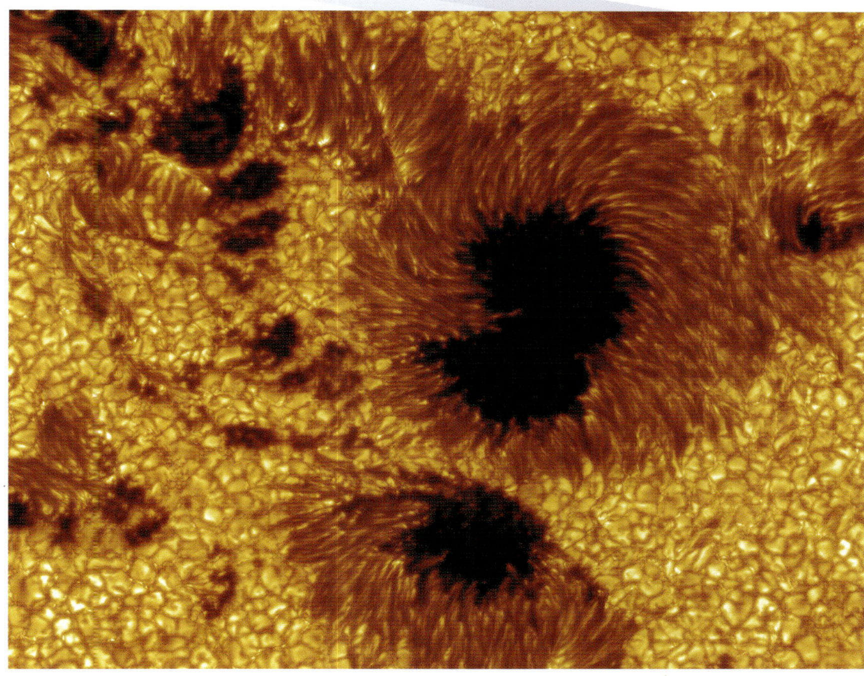

A close-up photograph of sunspots.

**Can you explain in your own words what solar flares, auroras, and sunspots are? How do they affect Earth?**

## *think about this*

Did you know that God designed the sun to get its power by little explosions that happen over and over again deep inside the sun? Something called thermonuclear (thur' moh new' klee ur) fusion is making all those little explosions. So next time someone asks you how the sun gets its power, tell them, "Thermonuclear fusion!" What is even more exciting is that thermonuclear fusion makes the sun brighter and brighter each year. Can you believe the sun is actually getting hotter and hotter as it gets brighter and brighter? It is!

Thermonuclear fusion shows that there was probably not any life on Earth billions, or even millions, of years ago. Since the sun is getting brighter and brighter each year, if we were to go back in time, we would see the sun getting dimmer and dimmer each year. In fact, if we were to go back billions of years, the sun would have been so dim, or faint, that it would not have provided enough warmth for life on Earth.

If temperatures on Earth were much cooler than they are now, there would be terrible consequences. It would be winter all the time. Even with the sun's current hot temperature, Antarctica is very, very cold. Scientists have discovered that the sun would have been many times cooler, actually more than 30% cooler, billions of years ago. Life, as we understand it, could not have survived on Earth if the sun were that cool because Earth would have had frozen land and surface water scattered about. This gives us evidence that life on Earth could be young, and not millions or billions of years old, as some suggest.

## The Color of God's Love

What color is the sun? When we look through special equipment or when the sun is setting in the sky, we see it as orange. However, the color of the sun is actually all the colors of the rainbow! So the sun is red, orange, yellow, green, blue, indigo, and violet. All those colors found in the sun's light make every single color in the whole world, so we have a very colorful sun. Let's spend some time learning about the light and colors that come from our wonderful sun.

Light always travels in a straight line. It does not bend or go around corners. You know this because when you put your hand in front of your face to shield your eyes from a bright light, like the sun, it casts a shadow across your face. The light does not bend and go around your hand into your face. The reason it is shady under a tree is because light travels in a straight line that does not bend around tree branches.

### *think about this*

C.S. Lewis was a very influential writer. He once said, "I believe in Christianity as I believe that the Sun has risen, not only because I see it, but because by it I see everything else" (1944). As you read through this lesson, you will begin to understand what he meant.

How do we get color from light? Light is a form of energy. Light energy travels out from the sun in a straight line. Although that line is going on a straight path, the line is curvy or wavy. In other words, light energy is wavy. Each color is a different size wave. Blue waves are short, as in the picture below. Yellow waves are long, as in the picture below that. It's strange to think of color as different kinds of waves, but that is what it really is.

BLUE LIGHT HAS SHORT WAVES.

YELLOW LIGHT HAS LONG WAVES.

How does it all work? Well, when light hits an object, many of the waves of light are absorbed into it. *Absorb* means to take in or soak up, the way a towel absorbs water. Almost everything absorbs at least some light waves. Some waves are absorbed, but some aren't. Instead, they bounce off. The waves that bounce off the object bounce up into your eye. Your eye sees the bounced light waves. When you see a yellow object, you are really seeing the yellow light waves that bounced off the object and hopped into your eye. The object has absorbed all other waves.

Your eyes and brain are what make the object appear yellow. God made your eyes and brain in such a marvelous way. Your eyes tell your brain what kind of light it is seeing, and your brain makes the color appear as it does. If there was no eye there to see the object, yellow light would still bounce off the object, but it wouldn't go into anyone's eye to see it as yellow. It would just bounce up until it hit an object that would absorb it. Here's something to think about: Is the yellow object still the color yellow if there is no eye there to see it? Hmmm . . . that is an interesting question. You decide.

All light waves are absorbed by the car except yellow light waves.

Why does the sun look orange at certain times of the day if its light is really all colors? Earth is covered with a layer of mist and gases called the **atmosphere** (at' muh sfear). When the sun's light waves travel through the atmosphere, the short-waved light bounces off the stuff in the atmosphere. When we look up at the sky, those bounced light waves hit our eyes, so our atmosphere appears the color of the bounced light waves. Remember that blue light has short waves. That is why the sky looks blue. All the blue light gets bounced off the stuff in the atmosphere, so the atmosphere looks blue.

Our thin, blue atmosphere.

If we look at the sun, we see light that has traveled straight from the sun to our eyes. In other words, we see the light that does not get bounced around. *If the sun's rays were not bouncing off stuff in our atmosphere, the sun would look white because all the colors combined make white light.* Sometimes the sun does look white when it is straight overhead. That's because the light is not passing through as much atmosphere when it is straight overhead, so not much light is bounced around. As a result, we see all colors of light coming from the sun, which makes the sun look white.

When you see the sun at sunset, the light waves travel through much more atmosphere, and most of the short-waved blue and green light has bounced off by the time it reaches your eye. This makes the sun look a very deep orange since those colors do not get bounced around in the atmosphere. We call this a beautiful sunset.

White light coming out of the sun is all the colors together except for black. What, then, is black? Well, when all the colors of light get absorbed into the same object, it makes the object appear black to the human eye. None of the colors bounced off, and that is why it is black. So a black object does not have any waves bouncing into your eye. It absorbs all the energy that the sun's light is pouring down on it. Nothing is bouncing up to make it a particular color. That is why black is not really a color. Color is what you see when something bounces into your eye.

All light waves are absorbed by the car.

If every single color from the rainbow bounced off the object and into your eye, what color would the object be? Well, think about it. All the colors from the rainbow make white light. So if the whole rainbow bounced into your eye, you would see the object as white!

White objects do not absorb much light energy because all colors of light bounce off them. Black objects absorb a great deal of light energy because they absorb all colors of light. That is fine inside when the light is coming from light bulbs. But when the light is coming from the sun outside, you will notice that black objects get very, very hot. They have a lot of energy sucked into them. White objects are not very hot. They reflect, or bounce off, all the sun's light. So when you are barefoot in a parking lot during the middle of the summer and you are waiting for your mom to unlock the car door, stand on the white dividing lines instead of the black pavement. They are much cooler.

**Do you understand why we see color?**
**Explaining what you have just learned to someone will help you remember it.**

## think about this

The Bible tells us we will not always need the sun as we do now. When the Lord returns and transforms the world back to its original perfection, God will be the light for us in the day and the night. His light will be all the light we need. Isaiah 60:19 says, *"No longer will you have the sun for light by day, Nor for brightness will the moon give you light; But you will have the LORD for an everlasting light, And your God for your glory."*

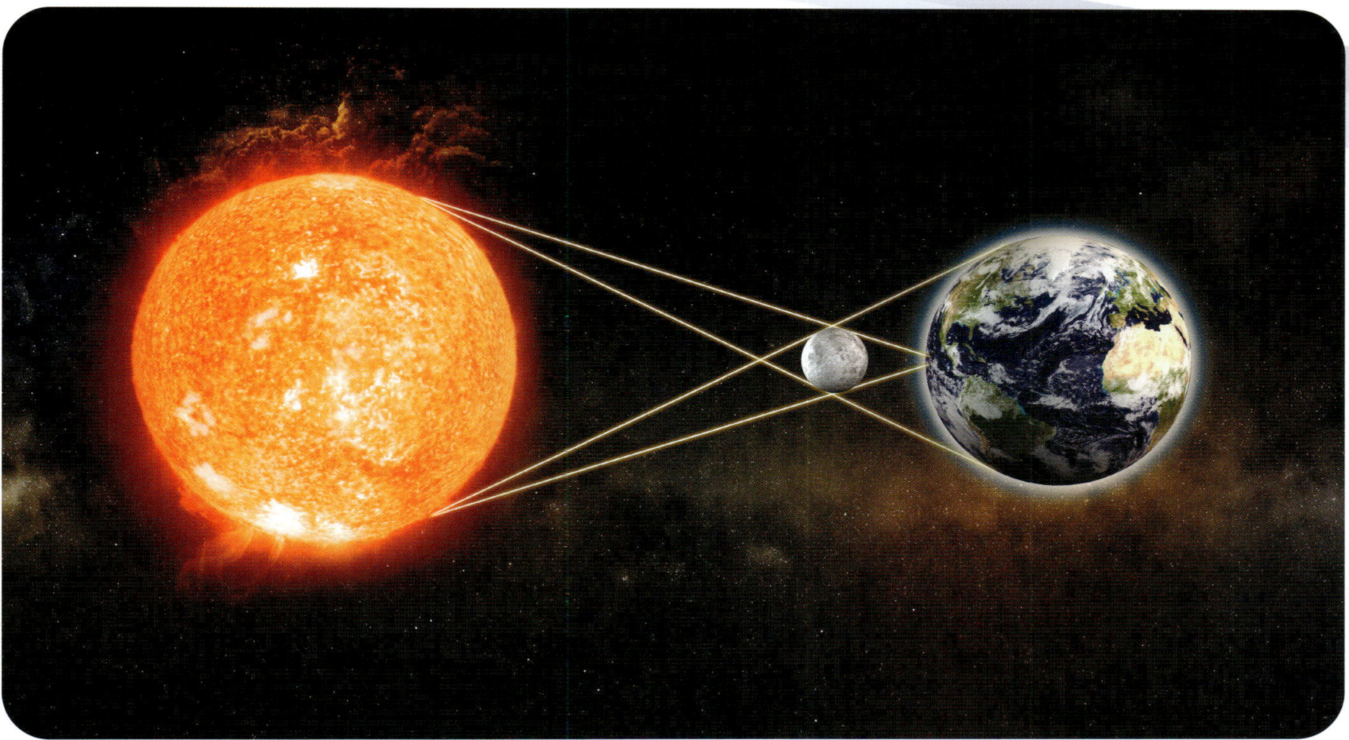

## Solar Eclipse

What would you think if you went outside one day, in the middle of the day, and there was darkness all around? You looked around for the sun, but it was completely gone. Something like this happens every once in a while. The sun seems to disappear in the middle of the day for a few minutes, and it seems like nighttime. Then the sun reappears. What makes the sun seem to disappear? It's the Moon! The sun gets hidden behind the Moon, and its light does not shine down upon us for a moment. It's called a **solar eclipse** (ee klips'). Every year at least 2 solar eclipses occur. Sometimes there are up to 5 in 1 year! You do not see this many because they are not all visible from where you live. Only certain parts of the world can see each eclipse. Most eclipses last for only a few minutes, but some have lasted as long as 7 minutes. When that happens, animals get confused and sometimes prepare to go to sleep, thinking it's nighttime.

In this solar eclipse, the dark circle in the middle is the moon. You can see that it is blocking out most of the light from the sun.

A solar eclipse happens when the Moon gets right in the path between Earth and the sun. Since the sun's light does not bend around the Moon, it casts the Moon's shadow across Earth. The Moon is much smaller than the sun, but since the Moon is so much closer to Earth, it appears bigger and blocks out most of the light from the sun for a few minutes. Do you remember activity 2.1? That is exactly what is happening here except that the Moon, not your finger, blocks out the light. Do you see how the Moon could make the sun seem to disappear?

When there is a total eclipse, the sun is completely hidden behind the Moon. During an **annular** (an' yuh ler) **eclipse**, the sun is directly behind the Moon, but a ring of sunlight can be seen around the blackened Moon. This happens because the Moon is sometimes farther away from Earth during an eclipse, and it cannot completely hide the sun.

A total eclipse of the sun.
You cannot see the sun.

An annular eclipse of the sun.
You can see the outer edge of the sun.

A partial eclipse of the sun.
You can see most of the sun.

Sometimes during an eclipse, the Moon is not directly between the sun and Earth but is only partially between them. These are called **partial eclipses**. *Partial* means part. During a partial eclipse, you can see a part of the Moon pass in front of the sun. It is still an amazing sight to see. Full and partial eclipses are astounding, but **you must look at them only through special eclipse-viewing glasses or a special eclipse-viewing box. Looking directly at the sun, even during an eclipse, could cause permanent eye damage and even blindness.**

The Moon is not a perfectly round ball. It has many gigantic holes, called craters, which give it an uneven surface. You can see these holes very clearly through a telescope. During an eclipse, little points of light reflect off these holes. We call these bright points of light **Bailey's Beads**. They look like little bright balls or beads.

A Closeup photograph of the Moon's craters.

Can you see the Bailey's Beads on the top of this picture?

# Activity 2.5
## Make a Solar Eclipse

### You will need:
- Adult supervision
- Flashlight
- Globe (or round ball)
- Small ball (smaller than the Globe)
- String

### You will do:
1. Tie string around the ball.
2. Place your globe on a table or floor.
3. Set your flashlight on a stack of books until it is shining directly on the center countries of your globe.
4. Turn off the lights.
5. Lower the smaller ball down between the flashlight and the globe, but make sure it is closer to the globe than it is to the flashlight. Move it until it is projecting a dark shadow on the globe, as shown in the picture.

### Discussion
Notice the dark circle that is projected onto the globe. That is called the **umbra** (uhm' bruh) of the eclipse. The people within that dark circle experience a total solar eclipse. The lighter shadow that surrounds the dark circle is called the **penumbra** (puh nuhm' bruh), and the people in that section experience a partial solar eclipse. Notice that the umbra and penumbra are not on most of the globe. The parts of the globe that do not have an umbra or penumbra do not experience any kind of eclipse. This is why a solar eclipse is not visible in all parts of the world. Though there are several eclipses a year, they are visible in only a few countries.

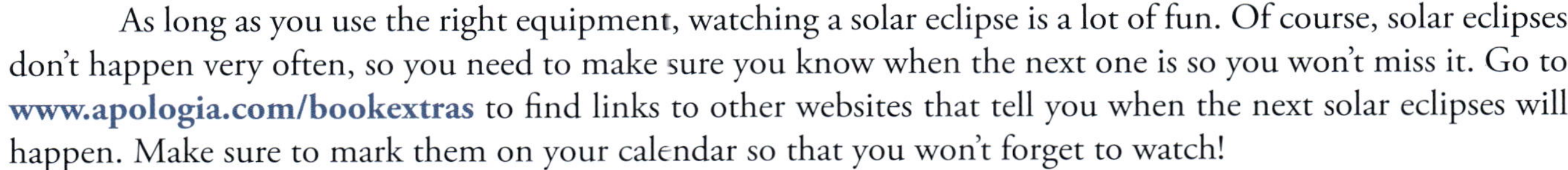

As long as you use the right equipment, watching a solar eclipse is a lot of fun. Of course, solar eclipses don't happen very often, so you need to make sure you know when the next one is so you won't miss it. Go to **www.apologia.com/bookextras** to find links to other websites that tell you when the next solar eclipses will happen. Make sure to mark them on your calendar so that you won't forget to watch!

**Use your own words to explain to someone how a solar eclipse happens.**

# Activity 2.6
## Pinhole Eclipse Viewing Box

You are going to make a pinhole eclipse viewing box that will help you to study the sun safely. You will not look at the sun directly through the pinhole. Instead, you will project an image of the sun onto a sheet of paper at the back of your box. The light from the sun will come through your pinhole and will shine an image of the sun onto the paper. You can use this to view the sun and, if the sunspots are big enough this year, you will be able to see them! Be sure to save your viewing box for when there is a solar eclipse. **Never look through the pinhole directly at the sun.** The pinhole will shine the shape of the sun onto the back of your box, and that's what you need to look at.

### You will need:

- Adult supervision
- Box
- Scissors
- White paper
- Pin or needle
- Tape
- Aluminum foil

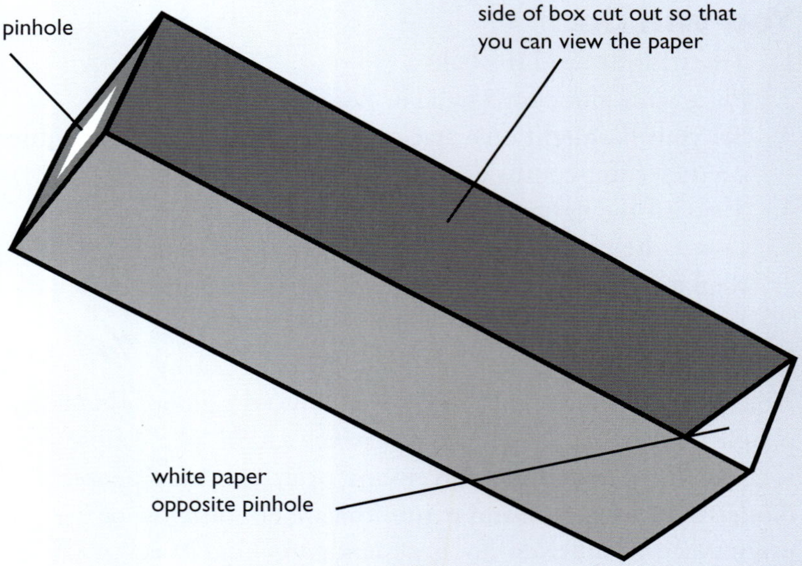

foil with pinhole

side of box cut out so that you can view the paper

white paper opposite pinhole

### Procedure:

1. Find a box. The length of the box is important. The longer the box, the bigger but fuzzier the final image. A shorter box gives you a smaller, clearer image. If you can't find a long box, you can tape 2 or more boxes together to make a longer one.
2. Use the scissors to cut a viewing hole in the side of the box. You can cut the whole side off the box or just cut a hole big enough so that you can see the back end of the box.
3. Use the scissors to cut a hole in the center of one end of the box. This will be the pinhole side.
4. Tape a piece of foil over the hole you just cut.
5. Use the pin to poke a small hole in the foil. You will not be looking through this pinhole; it is for the sun to shine through onto the other side of the box. A small hole will give you a sharp but dim image. A larger hole will give you a bright but fuzzy image.
6. Put a piece of white paper inside the back end of the box, opposite the pinhole.
7. Point the end of the box with the pinhole at the sun so that you see a round image on the paper at the other end. If you are having trouble pointing the box at the sun, look at the shadow of the box on the ground. Move the box so that the shadow looks like the end of the box. In other words, make sure that the sides of the box are not casting a shadow. The round spot of light you see on the paper is a pinhole image of the sun.
8. As you look at the image of the sun, see if you find dark spots within the image. Those are the sunspots! Make an illustration of them in your notebooking journal.

**(Continued on next page.)**

9. If you want a brighter image of the sun, you can remove the foil that you currently have and replace it with new foil. Then, poke a larger hole in the new foil. If you want a sharper (but dimmer) image of the sun, replace the foil with new foil and poke a smaller hole in it.
10. If you use your viewer during an eclipse, you can safely observe the Moon moving over the sun, blocking its light.

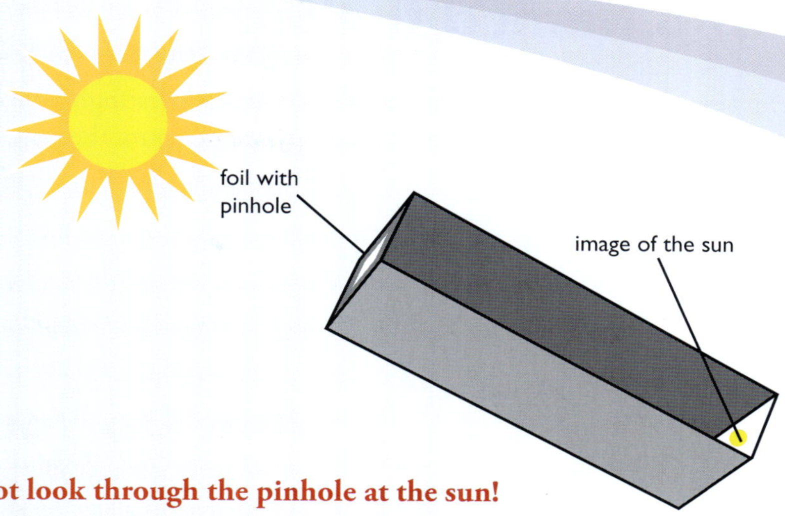


**Look only at the image on the paper! Do not look through the pinhole at the sun!**

You can see how big the SOHO spacecraft is when compared to these engineers.

## A Spacecraft to Study the Sun—SOHO

Remember reading about NASA in lesson 1? NASA is studying the sun with a spacecraft called SOHO that was launched in 1995. SOHO stands for the Solar and Heliospheric Observatory. SOHO studies what is happening inside the sun as well as events on the surface of the sun, such as the sun's solar flares. If you would like to build a paper model of SOHO, go to **www.apologia.com/bookextras** for instructions. If you make any models, consider hanging them from the ceiling with string. Wherever you keep them, be sure to take a picture for your notebook so you can remember all the fun you had while learning about the universe.

## Who Named the Sun

The Romans called the sun Sol, which is why our system of planets is called the Solar System. In modern English we now say sun. Interestingly, the ancient Greeks, called the sun Helios. This is why the heliocentric model puts the sun in the center of the solar system. One more interesting fact is that the sun has its own recognized symbol.

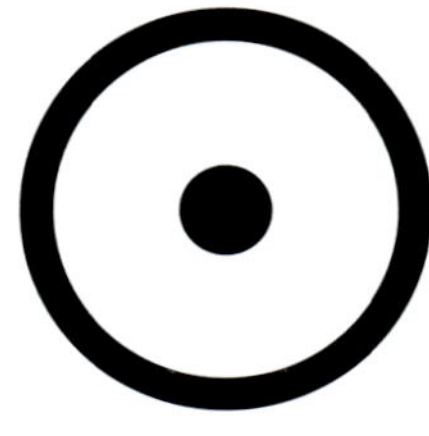
Astronomers use this symbol for the sun.

## What Do You Remember?

How many Earths would fit inside the sun? How many miles away is the sun? Can you explain the difference between revolving and rotating? Does the sun have any satellites? Explain what sunspots are. Do you remember why you see color? Can you explain what a solar eclipse is? What was your favorite part of this lesson?

LESSON 3
# MERCURY

## wisdom from above

As we begin our fascinating journey through the planets in our solar system, you will learn that the planets are all very different from each other.

*Great are the works of the Lord;*
*They are studied by all who delight in them.*

Psalm 111:2

# The Planet Closest to the Sun

Because Mercury is the closest planet to the sun, we will study it next. We do not know as much about Mercury as we do about many of the other planets in our solar system. Because it is so close to the sun, it is very hard to observe through telescopes. Two NASA spacecraft, *Mariner 10* and *Messenger*, have studied Mercury. Although *Mariner 10* and *Messenger* have collected much information about Mercury, it is still very little compared to what we know about many of the other planets. Of course, we do know *some* things about Mercury, and you will learn many of those things in this lesson.

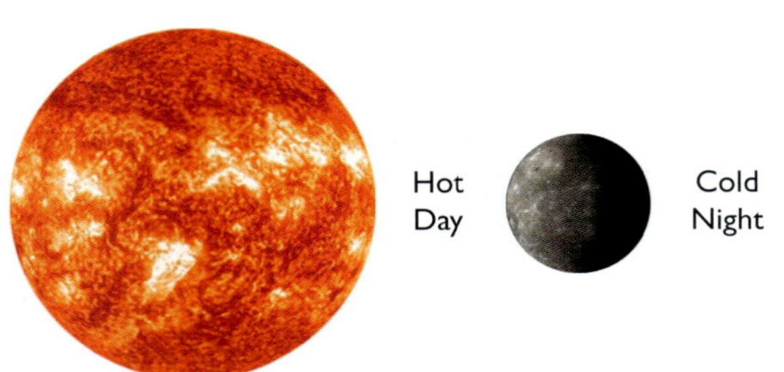


If you were standing on Mercury, the sun would seem gigantic because you would be closer to it. And boy, would it be hot and bright! During the day, when the sun is shining on Mercury, it gets sizzling at 800 °F (427 °C). That's hotter than an oven! You might think that since it's so close to the sun, it would never get cold there. You might think that it's always hot, but that is not so. Mercury cools down a lot at night because the sun does not shine on the side facing away from it. When it is night, Mercury gets colder than a freezer (-290 °F, -179 °C). This happens because Mercury does not have an atmosphere. As you learned in lesson 2, the atmosphere is the layer of mist, clouds, and gases that covers a planet and holds the heat in. This explains why Earth stays pretty warm at night. We would never feel very comfortable on Mercury. We would either be too hot or too cold. Weather is not the only reason we could not live on Mercury. Earth's atmosphere (we call it *air*) is made up of several gases and, as you probably know, one of those gases is the oxygen that we breathe to stay alive. Because Mercury does not have an atmosphere, it does not have oxygen and without oxygen, we could not breathe on Mercury without a spacesuit.

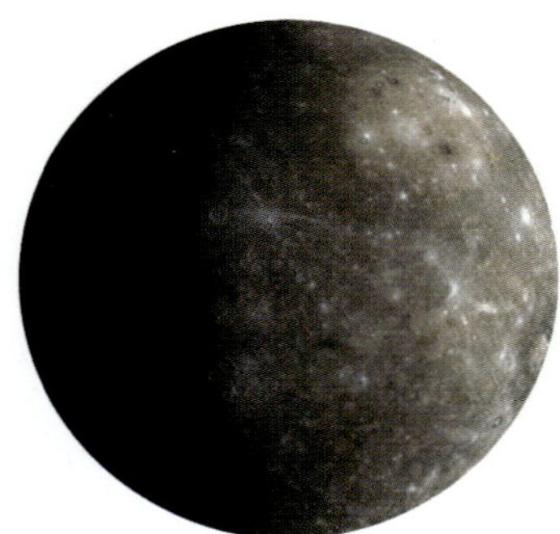

A picture of Mercury taken by *Messenger*. It resembles Earth's Moon.

Because Mercury doesn't have an atmosphere, the sky always looks dark. There are no air particles to scatter the light waves. Remember, the reason Earth's sky looks blue is because particles scatter the sun's blue light across the sky. Can you guess what color the sun would be to someone standing on Mercury? The sun would appear white all the time!

**Tell someone in your own words what you remember so far. Why is it so hard to study Mercury? What is the temperature like on the surface of Mercury?**

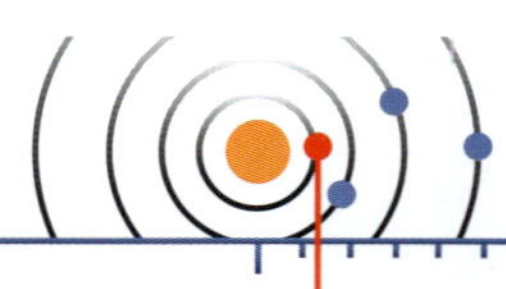

Mercury is 36 million miles away from the sun.

# Rotation and Revolution

Remember that a planet spins around in place while at the same time moving in a circle around the sun? The spinning in place is called *rotating*, and moving in a circle around the sun is called *revolving*. Planets rotate as they revolve. Of course, as one part of the planet is facing the sun, the opposite part is facing away from the sun. Earth takes 24 hours to rotate once. This means that if you are on the side of Earth that is facing the sun, you will be in the same position (facing the sun again) in 24 hours. We call this an *Earth day* because it is the length of time that a day lasts on Earth. Compared to Earth, Mercury doesn't spin fast at all. It spins very slowly. Mercury takes 59 Earth days to rotate one time. So if you were on Mercury, the bright, hot sun would shine on and on for many Earth days. Then, it would be cold and dark for many Earth days.

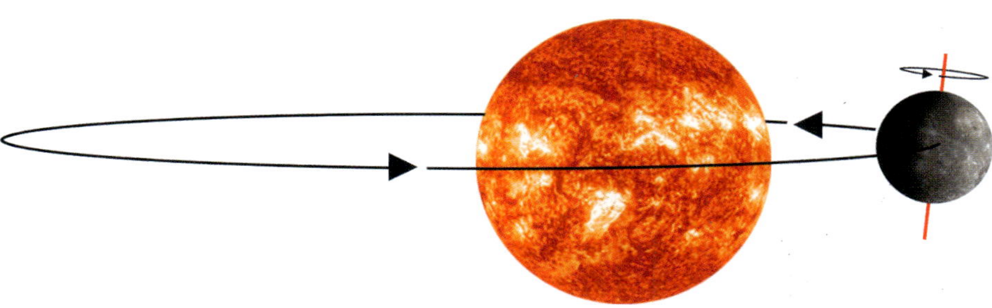

Because Mercury's orbit is elliptical, it is sometimes closer to the sun and sometimes farther away from the sun. In this drawing, it is closer to the sun when it is on the right (where it is drawn), and it is farther from the sun when it is on the left. Mercury's rotation makes the sun rise in the east and set in the west, just like on Earth.

We would miss the nighttime if we had to live a full day on Mercury, and we would miss the daylight if we had to spend a full night there. Imagine going to bed at night with the bright, hot sun shining through your window and waking up with that bright, hot sun still shining. Or, imagine it the other way. You wake up to pitch dark, spend the day in freezing darkness, and go to bed only to wake up to freezing cold darkness again for many Earth days! God certainly did not intend for people to live on Mercury. Only one planet was perfectly created for humans—Earth! Earth gets dark just about the time we feel tired, and about the time the sun is coming up, we are ready to wake up! It's not by accident that the Earth is perfectly suited for our sleep cycle. God made Earth's rotation just for us.

Even though Mercury is rotating very slowly, it's revolving around the sun very quickly. It's practically racing around the sun. As you learned in lesson 2, one trip around the sun is a full year for a planet. It only takes 88 Earth days for Mercury to make one trip around the sun. That's pretty fast. Mercury is the fastest planet in the solar system. It takes Earth 365 Earth days, to revolve around the sun. That means, in the time it takes to celebrate 1 birthday on Earth, you would celebrate 4 birthdays on Mercury! Think of all that yummy cake! Strangely, one Mercury day (one rotation) is only a little shorter than one Mercury year (one revolution around the sun). Remember, a day on Mercury is 59 Earth days, while a Mercury year is only 88 Earth days.

Mercury revolves around the sun in an oval pattern, not in a circle. We call it an **elliptical** (ee lip' tik uhl) orbit. If it traveled in a perfect circle around the sun it would always be the same distance away from the sun. An elliptical orbit makes it sometimes closer to the sun and sometimes farther away. How do you think it affects Mercury to be closer to the sun? How do you think it affects Mercury to be farther away?

Actually, none of the planets orbits the sun in a perfect circle. They all have elliptical orbits. However, most of the other planets (including Earth) have orbits that are almost circular. In other words, even though Earth is sometimes a little closer to the sun and sometimes a little farther away from the sun, the difference is very, very small. Because of this, we usually call the orbits of most of the other planets circles, even though they are actually elliptical. Mercury has one of the most elliptical orbits in our solar system.

## think about this

If you were standing on Mercury (in a spacesuit of course), the morning sun would appear to rise briefly, set, and then rise again. This is because Mercury has an elliptical orbit as well as rotates so slowly. At sunset, the sun would set, rise and then set again. Now that would be something to see!

**How long does it take Mercury to make one rotation?**
**How long does it take Mercury to make one revolution?**

# Features of the Planet Mercury

Mercury is small. It is much smaller than Earth. If Earth were the size of a baseball, Mercury would be the size of a golf ball. It's about the same size as our Moon. Mercury is the smallest planet in the solar system.

A size comparison of Earth and Mercury.

Mercury is a rocky planet. We call it a **terrestrial** (tuh res' tree uhl) planet. Terrestrial means *Earth-like*. Of course, this doesn't mean that a terrestrial planet is *a lot* like Earth. It just means that the planet is solid. In other words, you can stand on its surface. There are 4 terrestrial planets in our solar system (Mercury, Venus, Earth, and Mars). The other 4 are **gaseous** (gas' ee us) planets (Jupiter, Saturn, Uranus, and Neptune). As you might guess, this means they are made of gas. They are not solid; there is no ground to stand on. If you tried to land on a gaseous planet, you would sink through it.

A picture of Mercury's surface.

What is Mercury like? It's empty, like our Moon. It is also very dry. If you stepped on Mercury you would get dust all over you. Mercury has many craters on its surface. A crater is a large dent on the surface of the planet. It is a place where space rocks, called **asteroids** (as' tuh roids), crashed into Mercury, leaving big dents. If you have ever had a sandbox, it would be like dropping rocks into the sand and then picking them up to see the dents they left. Craters are like a scar on the surface of a planet.

The asteroids that astronomers believe fell out of the sky and crashed into Mercury could have come from a planet that exploded at one point in time. Many believe that there was once a planet between Mars and Jupiter that exploded and sent pieces flying into space. The planets and moons that did not have strong atmospheres to protect them would have received a lot of scarring. Earth's atmosphere burns up asteroids before they reach our planet. Mercury doesn't have this protection, so if thousands of gigantic rocks flew into Mercury, they would have caused huge pits on its surface.

Scientists believe the inside of Mercury has the same material that is inside Earth. This would mean that Mercury has an iron and nickel core. The core is the center of the planet, just like an apple's core is at its center.

## *think about this*

Mercury's largest crater (Caloris Basin) is as big as the state of Texas! If you stood on the edge and looked down into the crater, it would be a long, long way down to the bottom. What is even more amazing is that the shockwave from the collision was so powerful it created a hilly region the size of Germany and France on the opposite side of the planet!

# Activity 3.1
## Making Craters

### You will need:

- Adult supervision
- Newspaper
- Flour
- Several pebbles of different sizes

### You will do:

1. The pebbles you collected will represent asteroids in this activity.
2. Cover an area of the floor with newspaper.
3. In the center area of the newspaper, make a mound of flour.
4. Drop a pebble onto the flour and then carefully, trying not to disturb the shape of the crater, pick up the pebble.
5. Record the size and shape of the crater in your notebooking journal then repeat the process with your other pebbles. If you want, you can experiment further with different drop distances.

### Discussion

Do you see anything special about the crater shapes that you made? What kinds of shapes were made when you dropped pebbles that were not perfectly round? What did the craters look like? In real craters, the force of the impact causes an explosion (like a bomb), and the ground is equally thrown in all directions. You should see that even pebbles that are not perfectly round make craters that are almost perfectly round. Isn't that interesting?

**Tell someone in your own words what you remember so far. Do you remember what an asteroid is? What are craters? Can you describe the two types of planets that exist?**

## A Trip across the Sun

Because Mercury is closer to the sun than we are, it sometimes comes between Earth and the sun. As you learned in lesson 2, when the Moon comes between Earth and the sun, we call it a solar eclipse, and the Moon can actually block out the sun. Since Mercury is much farther away from us than the Moon, it cannot block out the sun. However, it can block out a tiny portion of the sun, forming a small, black dot. That's what the picture on the right shows. When astronomers see this, they say that Mercury is *transiting*, or passing over, the sun.

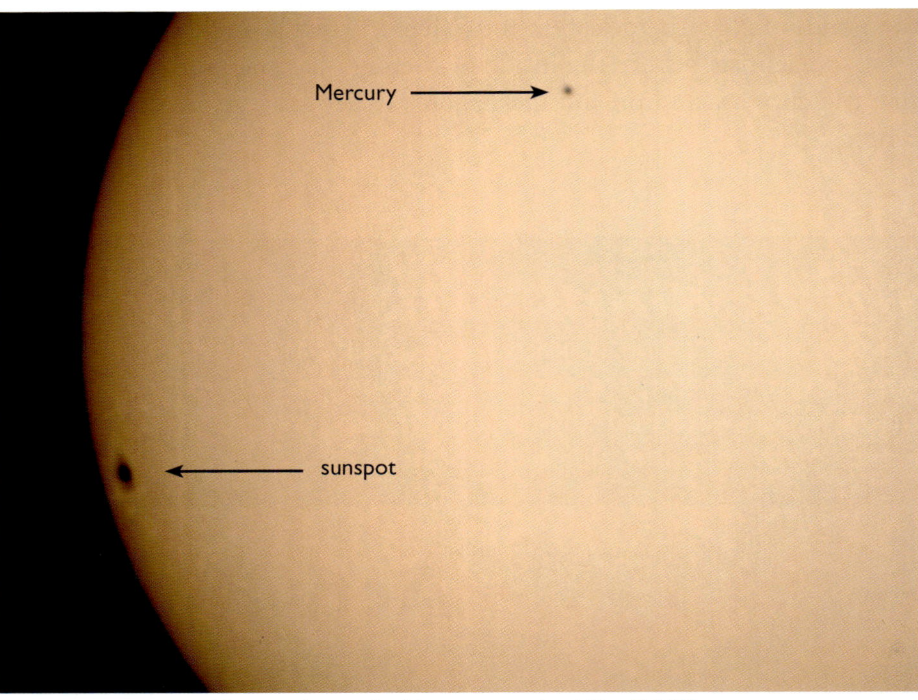

The sun during the Mercury transit of May 7, 2003.

Because Mercury is traveling around the sun, if you were to watch a transit of Mercury, you would see the black dot move across the sun. Scientists took several photographs of the sun during a recent Mercury transit, and compiled them into a **composite** (kahm pahz' it) image. Look at the trail of black dots in the picture. It shows the path Mercury took as it traveled across the sun.

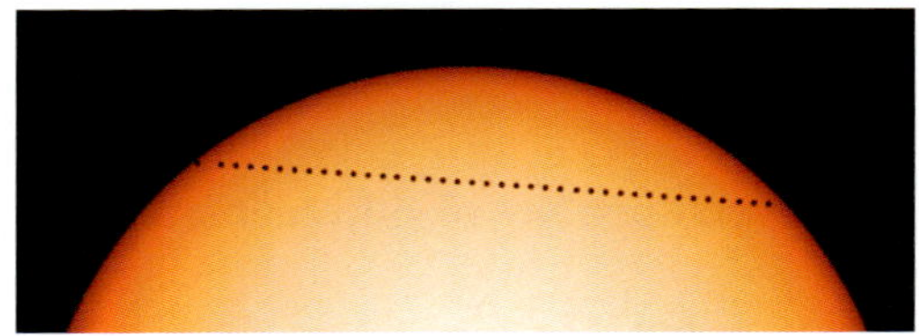
A composite photograph of the upper portion of the sun during the May 7, 2003 Mercury transit.

Astronomers use this symbol for Mercury.

## Who Named Mercury?

Sadly, the Romans who first named Mercury did not know God. The Romans worshipped idols and named the stars in the sky after these idols. The Romans noticed that Mercury moved very quickly through the sky. Remember, Mercury's orbit is so fast that it travels around the sun 4 times in just one Earth year! Because of this, the Romans named Mercury after their god of travel. When Jesus came, many Romans gave up their false gods, like Mercury, and became Christians.

Tell someone about the new facts you are learning.

## How to Find Mercury in the Sky

Did you know you can see Mercury without a telescope? You can see it either just before the sun comes

up in the morning or just after the sun goes down in the evening. If you look carefully toward the rising or setting sun, you may just spot this special planet very close to the horizon. Even though it shines like a star, there is really no light coming from Mercury. It looks like a star because it reflects the sun's light. Mercury is hard to find because it is often hidden from our view by the glare (bright rays of light) from the sun.

Even though Mercury is very hard to find in the sky, you can get some help finding Mercury by going to **www.apologia.com/bookextras**. The website has links to places that will show you where Mercury can be found.

## Spacecraft to Mercury

No man has ever gone to Mercury, but we have sent 2 spacecraft to Mercury to get information for us. As I told you at the beginning of the lesson, the 2 spacecraft were named *Mariner 10* and *Messenger*. No person was on either spacecraft, so we call them **unmanned spacecraft**. Even though there were no people on these spacecraft, there were many, many scientific instruments onboard. These instruments allowed the spacecraft to collect all kinds of interesting information.

An artist's idea of what *Mariner 10* looked like as it traveled through space.

*Mariner 10* was launched in November of 1973, and its mission was to take pictures of and collect information about Venus and Mercury. *Mariner 10* reached Venus in February of 1974. After scientists used *Mariner 10's* instruments to study Venus, they changed its course and sent it to Mercury. It reached Mercury in March of 1974. It was able to fly past Mercury again in September of 1974, and again in March of 1975.

# Math It

How old was your grandmother in 1974? Ask your grandmother what year she was born, then do a subtraction problem. Take the year your grandmother was born and subtract it from 1974. That will tell you how old she was when *Mariner 10* first visited Mercury. For example, if your grandmother was born in 1968, you would do the problem like this:

**1974**
**-1968**
**6 years old!**

If your grandmother was born after 1974, ask your grandfather what year he was born, or you can ask an older neighbor.

*Messenger* is another spacecraft that has gone to Mercury. In 2011, it made it all the way to Mercury. It was the first spacecraft to orbit Mercury. Because Mercury is so close to the sun, *Messenger* required a special shield to protect it from the heat of the sun. Can you see the shield on the illustration? On the side nearest to the sun, the temperature is over 500 °F (260 °C). Do you think it's warmer or cooler on the other side, shielded from the sun? It's cooler! It's about 100 °F (38 °C) there. *Messenger* has sent us a lot of information about Mercury. In fact, it has taken pictures of the whole surface of Mercury.

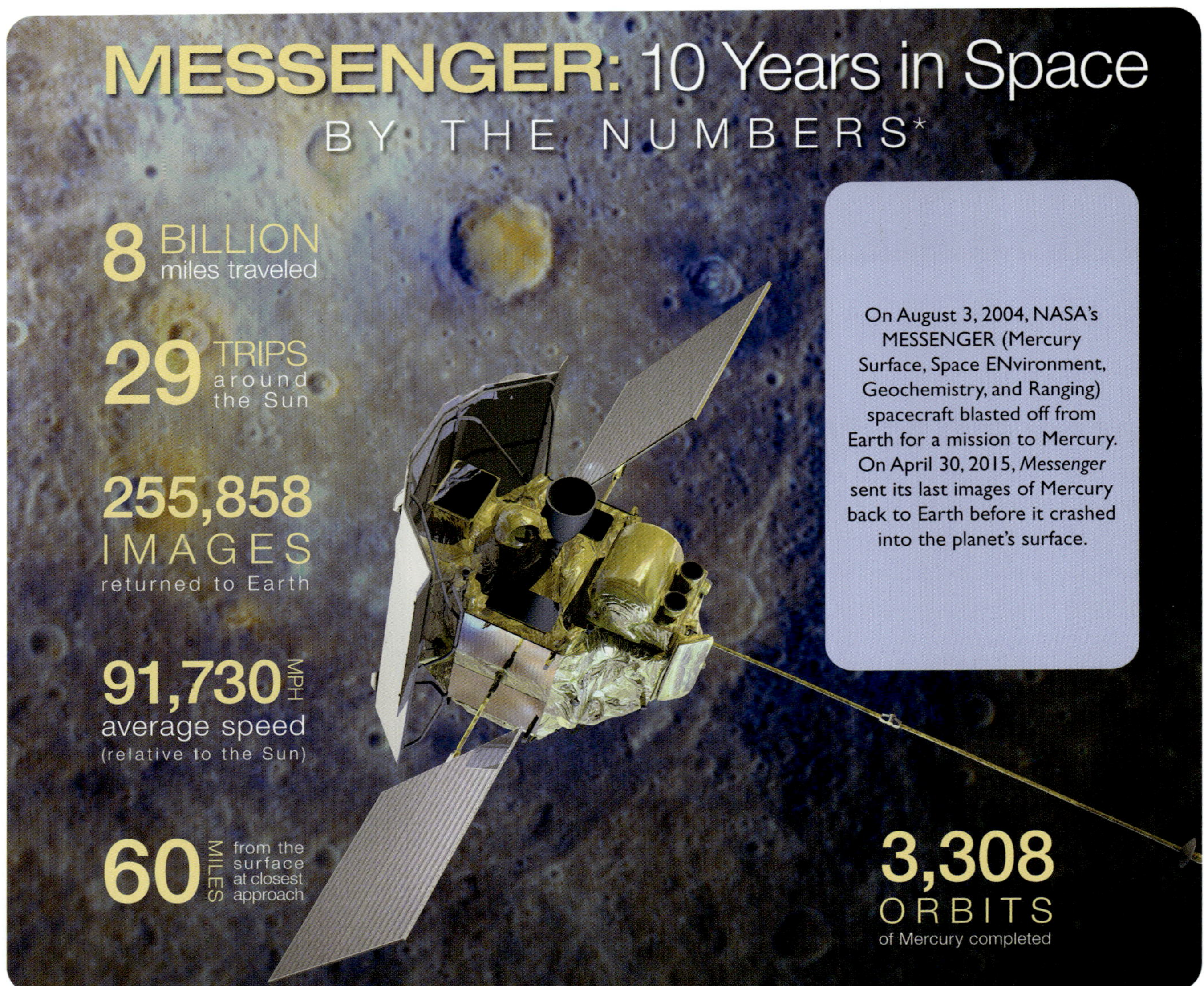


## think about this

"Messenger brought to light the intricacies of an intriguing world. The mission discovered a surface rich in diverse chemistry, and a bizarrely offset magnetic field. It photographed strange 'hollows' where material seems to have boiled away into space under the scorching sun and also uncovered deposits of water ice in the depths of polar craters where the sun never shines." - NASA

Two more spacecraft will arrive at Mercury in 2022. These spacecraft are the *Mercury Planetary Orbiter* (MPO) and the *Mercury Magnetospheric Orbiter* (MMO). These 2 spacecraft will teach us even more than we know today.

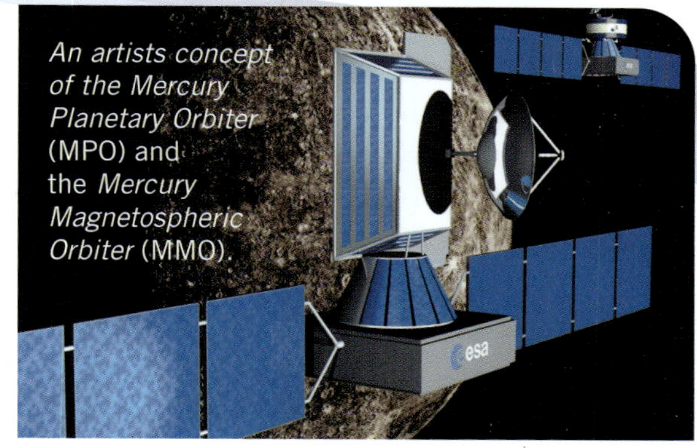

*An artists concept of the Mercury Planetary Orbiter* (MPO) and the *Mercury Magnetospheric Orbiter* (MMO).

# Activity 3.2

## Make a Model of Mercury

### You will need:

- Adult supervision
- Small bowl
- Marble or pebble
- Pencil
- Salt dough recipe

**SALT DOUGH RECIPE:**
1T flour
1 t salt
1 t water
1 drop blue food coloring
1 drop green food coloring
3 drops red food coloring

### You will do:

1. Mix your dough ingredients in your bowl.
2. Form the dough into a ball.
3. After you have made the ball of dough, you want to make it look like Mercury. To do this, you will make the craters on the ball. Look at the pictures below to see what your craters should look like:

**(Continued on next page.)**

4. You can use marbles or pebbles and pretend they are asteroids falling onto the planet, making the craters. Since the craters should be different sizes, you should also make some craters with a pencil. Use both sides of the pencil (the tip and the eraser)
5. When you are finished, place your model of Mercury somewhere to dry.

Example model of Mercury.

## What Do You Remember?

Is it hot or cold on Mercury? What does the sky look like from the surface of Mercury? Why? How long is a day (one rotation) on Mercury? How long is a year (one revolution around the sun)? Does Mercury orbit the sun in a circle or an oval? What is the shape of Mercury's orbit called? What kind of planet is Mercury: terrestrial or gaseous? What does the surface of Mercury look like? What are some reasons it might look like this? When is the best time of the year to see Mercury without a telescope? Why? What was your favorite part of this lesson?

LESSON 4

# VENUS

## *wisdom from above*

Our next planet, Venus is very different from Mercury. As we explore creation together you will be amazed at the wonders of our God and the diversity of our universe.

*How great are Your works, O LORD! Your thoughts are very deep.*

Psalm 92:5

Hot! Hot! Hot! The hottest planet in the whole solar system is not the planet closest to the sun. It's the one that is second closest. Venus is burning hot and stays hot all day and all night. It is about 870 °F (466 °C) on Venus. It also has numerous volcanoes that spew lava onto its surface. Do you know what **lava** is? It is rock that is so hot it has melted, like butter melts in a pot on a stove. Lava, however, is much, much hotter than that. It is so hot it would melt the pot holding the butter, turning the pot into liquid! Lava comes out of a volcano when it erupts. When it is inside a volcano, lava is called **molten rock** (melted rock).

Venus has thousands of volcanoes all over it. They blow up every now and then, spouting and spraying hot, fiery lava. Many of the parts of Venus that are not covered with hot lava are covered with places where lava once flowed and then cooled down enough to form hard rocks. Even though these hard rocks are no longer lava, they are still very hot because everything on Venus is hot. Venus's whole surface is affected by these volcanic eruptions. When a volcano erupts on Earth, it wipes out every living thing it touches. Every tree, house, plant or animal that it touches becomes nothing but ash when the lava dries. When a volcano erupts, for years afterward, people are still trying to clean up all the ash that blows around the area.

Computer simulated global view of Venus.

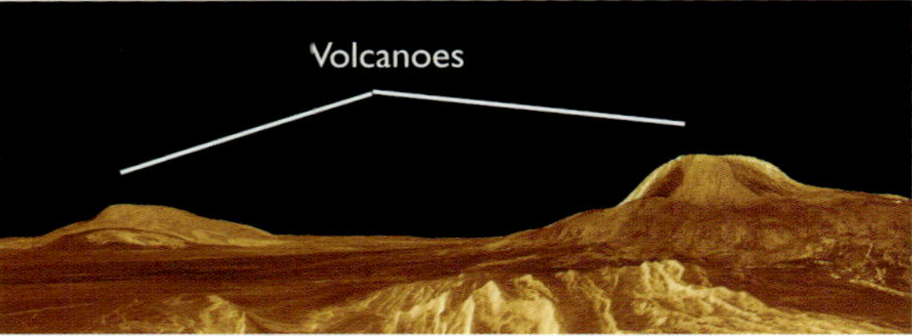


The surface of Venus. There are 2 volcanoes in the background, and the lighter-colored rocks are rocks that were once lava but have since cooled to form hard rocks.

Venus is 67 million miles away from the sun.

Do you have a fireplace? If you have real logs in your fireplace, you have seen the ashes that are left when they have burned up. Venus probably doesn't have ashes all over it, because there is nothing living on Venus, so there is nothing to burn up and turn into ash.

With all these rocks all over Venus, do you think it is a terrestrial planet or a gaseous planet? Remember that *terrestrial* means Earth-like. Yes, Venus is a terrestrial planet. If a planet has a rocky or earthy surface, it is terrestrial because you can stand on its surface, just like you can stand on Earth.

## *think about this*

When you read that Venus is very hot, you probably don't think this hot: "On Venus you could cook a 16-inch pepperoni pizza in seven seconds, just by holding it out to the air." (*Death by Black Hole*, p. 80)

# Activity 4.1

## Make Some *Lava*

### You will need:

- Adult supervision
- 1 T butter
- 1 T flour
- Small plate
- Thimble

### You will do:

1. Melt the butter.
2. Put the thimble upside down in the middle of the small plate.
3. Sprinkle flour around the thimble so that you have a nice layer of flour surrounding it.
4. Slowly and carefully pour the butter onto the center of the thimble. Watch it roll down and into the flour. As it does so, you should see it form streams of melted butter in the flour.
5. Let the plate sit out for a few hours. Do not disturb it.
6. Come back later and look at what has happened.

### Discussion

In this activity, the butter represents rock. It begins as a solid, but when heated, it melts, turning into a liquid. Rocks do the same thing. When heated in the hot insides of a planet, rocks melt. When you poured the liquid butter over the thimble, it was like lava pouring down the sides of a volcano. It cut little streams out of the flour, going where it wanted to go. After you let the plate sit out for a while, what happened? The butter cooled and became solid again. After lava comes out of a volcano and sits for a while, it hardens into rock.

**Tell someone in your own words what you know about Venus. What is the temperature like on Venus? Why isn't there any ash on Venus?**

## Too Much Atmosphere

You might be wondering why Venus is hotter than Mercury, which is the planet right next to the sun. There is a very good reason. Do you remember what the word *atmosphere* means? The atmosphere is made up of all the gases that surround a planet. Venus has a very heavy atmosphere, with thick, heavy clouds swirling around it. Those clouds are similar to our clouds in one way—they create lightning. Have you ever seen lightning strike? If you were on Venus, you'd see it a lot.

The clouds and atmosphere trap the heat that comes from the sun, and the heat cannot escape Venus and go out into space. The sun's heat comes in, but it doesn't go out very easily. This keeps Venus very hot. The clouds on Venus are made of different stuff from the clouds on Earth. They are heat-trapping clouds made of sulfuric (suhl fyur' ik) acid. Sulfuric acid is not good to breathe.

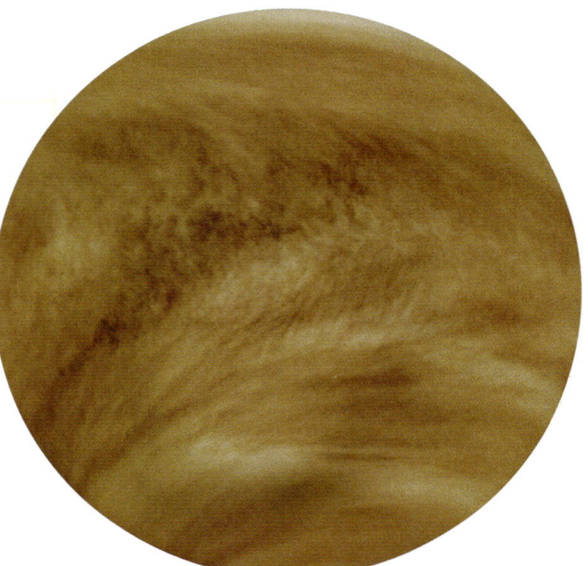

A picture of the clouds that constantly surround Venus.

When your mother is boiling water in a pan on the stove, it will boil faster if she puts a lid on the pot. The lid keeps the heat from escaping, trapping it inside the pot. This makes it hotter and hotter inside the pot so that the water will boil more quickly. The clouds on Venus are similar to the lid on the pot; they surround the planet, keeping the heat from escaping into space.

The clouds and atmosphere that trap heat on Venus are always moving across the planet. They move quickly (over 220 miles per hour) and are always circling the planet, heating up everything they pass over. So, the side of Venus that faces away from the sun isn't cooler than the side that faces the sun. Remember, this is not what it's like on Mercury. On Mercury, the side of the planet that faces away from the sun gets very cold because there is no atmosphere and no clouds to keep it warm. On Venus, the side that faces away from the sun is kept warm by the atmosphere and clouds, so it never has a chance to cool off. Because of this, it's just as hot at night on Venus as it is during the day.

An artist's drawing of what lightning on Venus might look like.

**Take a moment to use your own words to describe the atmosphere on Venus.**

# Rotation and Revolution

Venus rotates slowly. Do you remember what *rotate* means? It is the circular movement that makes a certain part of a planet face the sun and then turn away from it, changing day into night. Venus rotates even more slowly than Mercury. It takes 243 Earth days for Venus to make a full turn. Because of this, daytime on Venus lasts just over 121 Earth days and nighttime lasts just over 121 Earth days.

Venus not only rotates slowly, it rotates in the opposite direction of Earth. When we see the sun come up early in the morning, we know we are looking east because the sun always rises in the east. When we see the sun going down, we are looking west because the sun sets in the west. But on Venus, the sun rises in the west and sets in the east! Remember, the reason the sun seems to rise and set is that the planet is rotating. The sun is not moving; the planet is turning toward and then away. Since Venus rotates opposite of Earth, the sunrise and sunset are opposite as well.

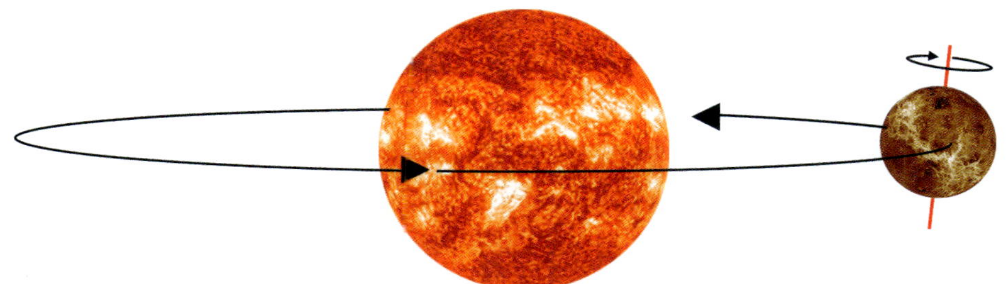

Venus rotates so that the sun rises in the west and sets in the east.

Believe it or not, a year on Venus is shorter than a day on Venus! This means that Venus orbits around the sun faster than it rotates! You see, it takes Venus only 225 Earth days to orbit around the sun, but it takes 243 Earth days for Venus to make a complete rotation. If you were on Venus, then, by the time day turned into night and back into day again, *more than a year* would have passed! This is not true for any other planet.

Now I want you to remember how long a year is on Mercury. Now compare that to the length of a year on Venus. Which is longer? A year on Venus is longer than a year on Mercury. It turns out that the farther a planet is from the sun, the slower it travels around the sun. Because of this, the farther a planet is from the sun, the longer its year will be. Since Mercury is closest to the sun, its year is very short. Can you guess which planet has the longest year? Think back to what you learned in lesson 1. Which planet is farthest from the sun? Neptune. That means Neptune has the longest year. You will learn how long Neptune's year is later on in the course.

**Can you explain in your own words all that you have learned about Venus so far?**

# Not a Twin

Venus is about the same size as Earth. Years ago, when looking at Venus through a telescope, Venus and Earth seemed similiar. Scientists said that Venus was Earth's twin. Many thought there might even be dinosaurs or other creatures roaming on Venus. Looks can certainly be deceiving! Today, with further study made possible by unmanned spacecraft, we know that Venus is very different from Earth. No plants or animals could live there; Venus is too close to the sun and too hot.

This is a size comparison of Venus (left) and Earth (right).

On Earth, the sun sets in the west. Because Venus rotates opposite the direction Earth rotates, the sun sets in the east on Venus.

## *think about this*

Although Venus isn't Earth's twin, there are some similarities.

Scotland is a place on Earth that is very mountainous (known as highlands). Venus has two such regions, but they are very large.

Istar Terra in the upper northern/polar area of Venus is about the size of Australia.

Aphrodite Terra at Venus's equator is about the size of South America.

Also, the highest mountain on Venus is called Maxwell Montes and is comparable to the highest mountain on Earth, Mount Everest. How might these mountains be different? Remember Venus is very hot. Mount Everest (on Earth) has snow at its peak. There is never, ever any snow on Venus!

# The Phases of Venus

If you look at Venus through a telescope, you will see that it appears to change shape from day to day. Galileo discovered this. Do you remember who Galileo was? He was the man who first began studying space with a telescope. He noticed that when he looked at Venus with his telescope, sometimes it looked like a curved sliver, called a **crescent** (kres' unt), sometimes it looked like half of a disk, and sometimes it looked like a full disk. At other times, it could not be seen at all. He called these shapes the **phases** of Venus, and he correctly concluded from this evidence that Venus orbited the sun.

In the drawing on the next page, you can see what Venus looks like to someone on Earth at different spots in its path around the sun. Remember, the planets do not shine with their own light. They must reflect the light of the sun in order to be seen. If Venus is between Earth and the sun, it reflects light back to the sun, not to Earth. During these times, Venus is completely dark to us and is not visible in the night sky.

The phases of Venus when viewed from Earth with a telescope.

When Venus is behind the sun, the side facing the sun reflects light back to Earth. Because we see an entire side of the planet, it looks like a bright disk. When Venus is to one side of the sun only about half of the planet can reflect light back to Earth, so we see the planet as half of a disk. At other points in its orbit around the sun, only a sliver of Venus can reflect light back to Earth, so we see Venus as a thin crescent. At other points in its orbit, all but a sliver of the planet can reflect light back to Earth, so we see most of Venus, with just a sliver removed. Of course, Venus is so far away that it just looks like a point of light to the naked eye. However, if you look at Venus through a telescope, you will see these phases quite clearly.

Look at the illustration of the phases of Venus again. Does that look familiar to you? It should. If you have spent much time looking at the Moon, you will see that its shape seems to change, much like what is shown in the drawing. That's because the Moon goes through phases, just like Venus. We will talk about the phases of the Moon in detail later on in this course.

## Who Named Venus

Because Venus is the second brightest object in the night sky (our Moon is brighter), ancient astronomers thought that it was very beautiful. The planet Venus is named for the Roman goddess of love and beauty.

### *think about this*

You know that Venus is not a star, but rather a very bright planet because it is closer to the sun. But, did you know that Venus is so bright it can cast shadows? The next time you are out on a moonless night, see if you can see your shadow by the light of Venus.

Astronomers use this symbol for Venus.

# Finding Venus in the Sky

Because Venus is close to the sun, we can only see it for a while before the sun rises and for a while after the sun sets. Of course, since Venus goes through phases, we won't always be able to see it in the night sky. When it is visible, however, Venus is the brightest thing in the night sky except for the Moon. It is often called the *morning star* or the *evening star* because it shines brilliantly after sunset or before sunrise.

On a very clear morning or evening, look toward the early rising or setting sun. You might see a bright point of light that looks like a star. That's Venus. Of course, you will know that it's not a star at all. It's a burning hot planet with lava and heat-trapping clouds made of sulfuric acid swirling madly around it. If you need help finding Venus, go to **www.apologia.com/bookextras**. It will tell you where to look for Venus and whether or not Venus is currently visible.

**Take a moment to use your own words to explain what you have learned so far.**

# Spacecraft to Venus

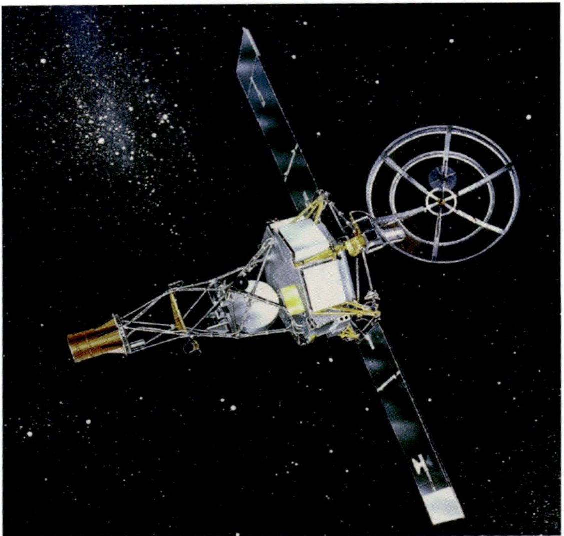

An artist's idea of what *Mariner 2* looked like as it flew through space to visit Venus.

Do you remember how many spacecraft have visited the planet Mercury? There have been 2. However, 26 spacecraft have visited Venus! All of these spacecraft have been unmanned, but they were equipped with lots of scientific instruments, so we know a lot more about Venus than we do about Mercury! The first NASA spacecraft to visit Venus was *Mariner 2*, way back in 1962. Do you remember the year your grandmother was born? You were supposed to learn that in lesson 3. If you don't remember, ask her again. If she was born before 1962, do the same kind of subtraction that you did in lesson 3 to find out how old she was when *Mariner 2* launched. If she was born after 1962, she had not even been born when *Mariner 2* visited Venus!

Now remember, Venus is very hot and is covered with sulfuric acid clouds. Most of the spacecraft that visited Venus never tried to land on its surface because the conditions on the planet would destroy them. Instead, most of the spacecraft simply orbited the planet, using their scientific instruments to gather as much information as possible. That was actually a very hard thing to do. Why? It is hard to see through the clouds that surround Venus, which means it is very hard to see the surface. Of course, scientists wanted to learn as much as possible about the surface of Venus, so they used **radar** (ray' dar).

# Understanding Radar

What is radar? Radar sends out signals and then waits for those signals to bounce off of something and come back. It measures the time it takes for the signal to return and uses that information to determine the distance of the object. In the case of the spacecraft orbiting Venus, radar units sent thousands of signals to the surface of the planet.

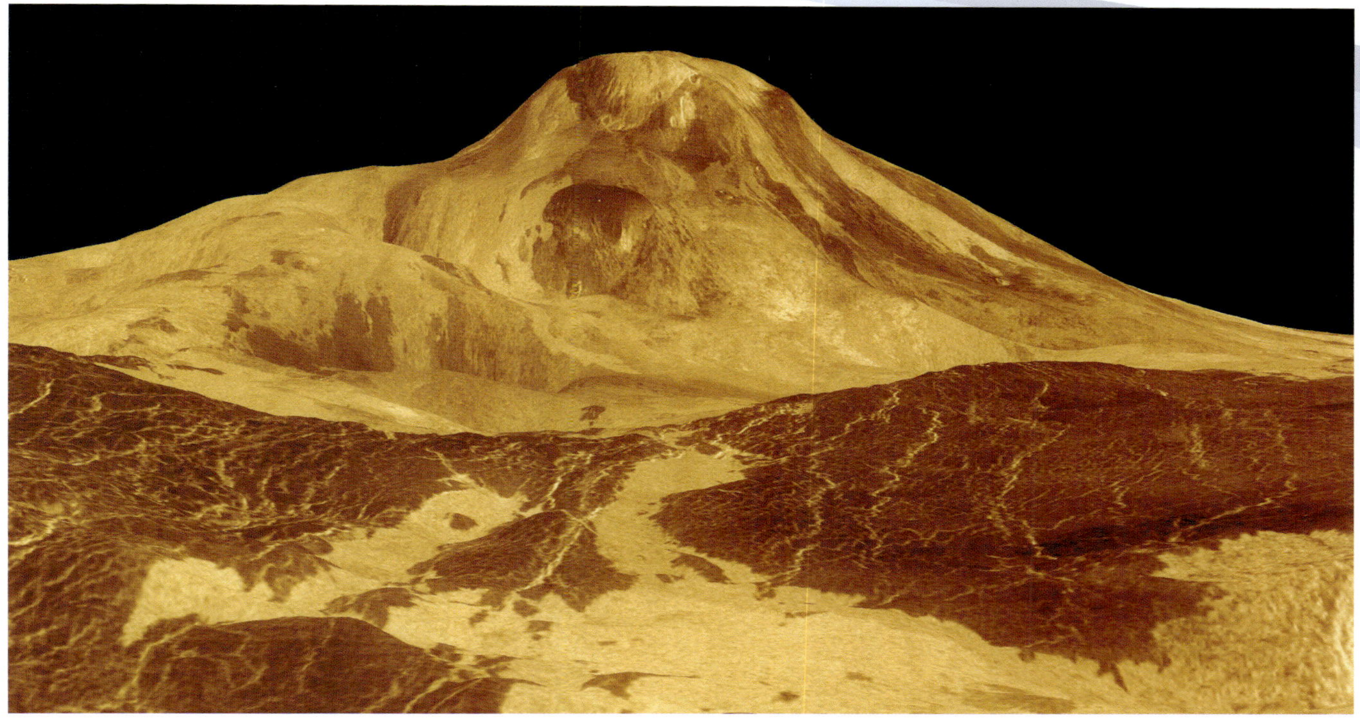
A volcano and the valley below it on the surface of Venus. The image was made using radar.

What good does that do? Think about sending a radar signal to one spot on the planet, and then sending another radar signal just to the right of the first. Suppose the radar said that the first signal traveled 500 miles and that the second signal traveled 505 miles. What would that tell you? It would tell you that the second spot on the planet is 5 miles lower than the first spot because the signal traveled 5 miles farther before bouncing back. In other words, there is a cliff there that is 5 miles high! If used carefully, radar can tell what the surface of the planet looks like, even though hidden by thick clouds. The activity at the end of this lesson gives you practice working with something similar to radar.

Although many of the spacecraft that visited Venus did not land on the surface of the planet, some did. The temperature and conditions of the planet were hard on the spacecraft, and some were destroyed either as they were in the process of landing or once they had landed. Even though these spacecraft were destroyed, scientists learned some important things. For example, a Soviet spacecraft called *Venera 13* was launched in 1981 and was able to land on the surface. Although it stopped working shortly after it landed, it was able to send some photographs back to Earth. One of those photographs is shown to the left. The picture you see on top is what the spacecraft took. The picture on the bottom is the same, but the effects of Venus's atmosphere are removed. In other words, this is what the picture would have looked like if Venus had no atmosphere. The difference between the pictures shows you how Venus's atmosphere colors everything on its surface.

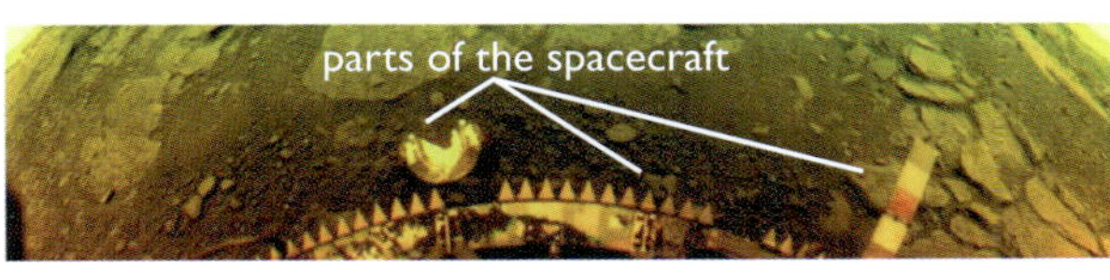

Views of a photo sent back from the surface of Venus.

# Activity 4.2

## Learn How Radar is Used

Because a thick layer of clouds covers Venus, spacecraft cannot see the surface. However, radar can reveal what the surface looks like. In this project, you will determine what the inside of a box looks like, even though you cannot see into it.

### You will need:

- Adult supervision
- Small box
- Strong paper towel (to cover the top and sides of the box; it should be thick enough that you cannot see through it and it won't tear with multiple holes poked)
- Tape
- Paper torn into different sizes
- Glue mixture
- Bamboo skewer (or a long, skinny stick with a point)
- Markers, crayons or colored pencils
- Ruler (or measuring tape)

**Glue Mixture Recipe:**

1/4 c Water
1/4 c Flour

### You will do:

1. Ask your parent to dampen the torn paper with the glue mixture and form different shapes, such as hills and valleys, that will fit inside your box and represent your planet's surface. These shapes can be air dried overnight or put into an oven at a low temperature, such as 170 °F for 30 minutes to dry.
2. When the shapes are dry, have your parent use tape to fasten them into the bottom of the box, leaving some spaces flat. You are not supposed to see into the box.
3. Your parent needs to place the paper towel over the box and secure it in place with the tape.

4. Look at what you have. You know that there is something in the box, but you can't tell what it is because the paper towel hides it. However, by using the *radar* in this experiment, you will be able to get a good idea of what is in there.
5. Begin by marking a grid on the cloth that is covering the top of your box. Do this by making rows and lines in both directions. The more lines you have, the more accurate your exploration of the box. Number each box on your grid, beginning with 1.

| | | | | | | | | | |
|---|---|---|---|---|---|---|---|---|---|
| 1 | 2 | 3 | 4 | 5 | 6 | 7 | 8 | 9 | 10 |
| 11 | 12 | 13 | 14 | 15 | 16 | 17 | 18 | 19 | 20 |
| 21 | 22 | 23 | 24 | 25 | 26 | 27 | 28 | 29 | 30 |
| 31 | 32 | 33 | 34 | 35 | 36 | 37 | 38 | 39 | 40 |
| 41 | 42 | 43 | 44 | 45 | 46 | 47 | 48 | 49 | 50 |
| 51 | 52 | 53 | 54 | 55 | 56 | 57 | 58 | 59 | 60 |

6. Copy the grid on the paper towel onto a blank sheet of paper. This is where you will record your data.
7. Measure the height of your box and use your markers to make equal increments of different colors. For instance, the first centimeter from the skewer's point is blue, the second is red, the third is yellow, and so forth.
8. Begin by sticking the bamboo skewer though the first square on the grid. Once it goes through the paper towel, push it down gently until you feel it touching something inside the box.
9. Look at the skewer to see the color that is closest to the paper towel. Fill in that color on the first box of your paper grid.
10. Continue to do this for each box on the grid.
11. Look at your paper grid. What areas tell you that there are mountains on your planet's terrain? What areas tell you that there are valleys? Are there any flat surfaces where you could land a spacecraft?
12. Carefully remove the paper towel from your box and see if your data was correct.

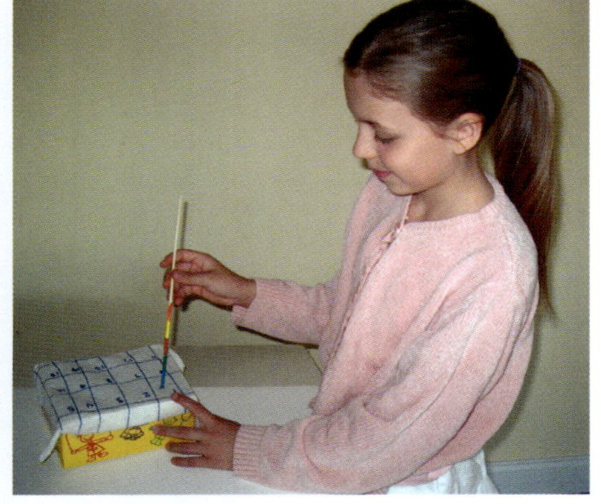

## Discussion

Think about what you have done here. Without seeing what was in the box, you were able to determine the hills and valleys that were in the box, and you were able to find a flat spot to land a spacecraft. This is how scientists use radar!

# What Do You Remember?

What would it feel like on Venus? What is the atmosphere like on Venus? What is special about the rotation of Venus? Why did astronomers think Venus was Earth's twin? Why does Venus go through phases? Have many spacecraft have visited Venus? Since we can't see through the thick clouds covering Venus, how do we know what the planet's surface looks like? What was your favorite part of the lesson?

LESSON 5

# EARTH

## *wisdom from above*

The Bible tells us that God made Earth. Genesis 1:1 says, "In the beginning God created the heavens and the earth." Remember we talked about how God makes everything for His glory? Well, if the beautiful heavens declare His glory, Earth must be screaming His glory because it is the most glorious of all planets. The Bible also tells us how loved we are:

*"The heavens are the heavens of the LORD, But the earth He has given to the sons of men."*

Psalm 115:16

# Perfect Design by a Perfect Designer

Now we get to take a closer look at the most fabulous planet in the entire solar system! I know, I know, Earth may seem a bit boring since you see it every day and already know a lot about it. But it's actually the most amazing planet in our whole solar system! God's fingerprints are all over Earth. When we look at the details that make Earth so perfect for us, it's obvious that only a very wise and wonderful God could have made it. Our planet is so special, so perfectly suited for people, plants, and animals, that it would be impossible for it to have happened by accident. God created Earth in such a way that if any part were missing, life could not exist. This means that Earth could be the only planet in the whole universe that has any life on it.

# Perfect Distance

Earth is in the zone! There's a small zone around the sun where life can exist. God placed Earth right in that zone—the most perfect place in the entire universe for us. We're not too far away or too close to the sun. Earth has plenty of water, which is the main ingredient for life to exist. Water is everywhere on Earth. There is even water inside of you!

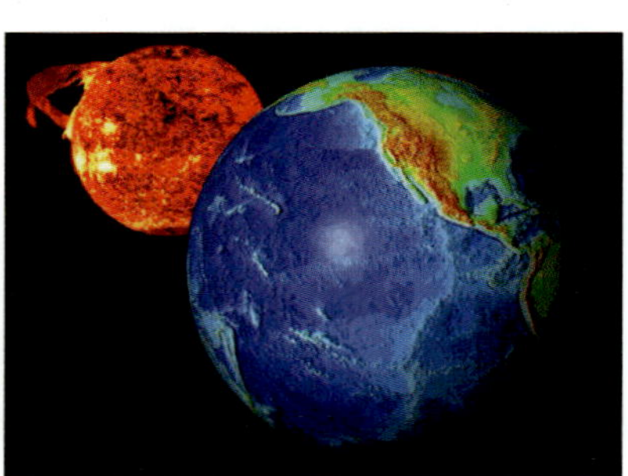

A model of Earth in front of a picture of the sun.

Do you know that if we were closer to the sun, the oceans would disappear? They would just dry up! The atmosphere covering Earth would also be destroyed, and the harsh rays from the sun would destroy every living thing on Earth. We could not survive on this planet if it were much closer to the sun.

If we were farther away from the sun, it would cause other terrible problems. The water would freeze, and we would have frigid, icy weather every single day of the year. Spring would never bring new plants, and the animals would eventually die of starvation. God placed us exactly where we need to be. He made Earth a perfect place for people to live!

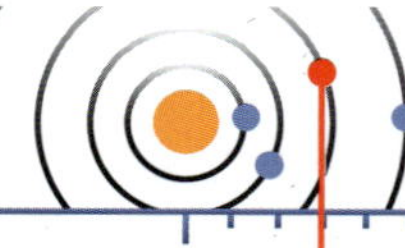

Earth is 93 million miles away from the sun.

# Perfect Mass

Do you have a baseball and a tennis ball around the house? If so, go get them. Hold the tennis ball in one hand and the baseball in the other. These 2 balls are almost the same size, but as you hold them, you should notice that the baseball is heavier than the tennis ball. Why is that? If they are pretty much the same size, why is one so much heavier than the other? The answer is that there is more **matter** in the baseball than there is in the tennis ball.

What is matter? It is the stuff that makes up everything around you. Everything you can touch or smell (including yourself) is made of matter. The more matter that is in something, the heavier it is. Since the baseball is heavier than the tennis ball, the baseball must have more matter in it. Now remember, these 2 balls are pretty much the same size. So, if the baseball has more matter in it, what does that tell you? It tells you that matter is packed into the baseball more tightly than matter is packed into the tennis ball. Think about putting toys into a box. If you put a ball and a squirt gun into the box, it wouldn't be very hard to lift. However, if you took all of your toys and stuffed them into the box, it would be a lot heavier, wouldn't it? Why would the box be heavier? It didn't change size. It just had more things packed into it. In the same way, the baseball has more matter packed into it than the tennis ball, so it is heavier than the tennis ball, even though it is close to the same size.

This balance tilts toward the baseball because the baseball has more mass than the tennis ball.

Scientists have come up with a way to calculate **mass**. Mass is a measure of how much matter is in something. If something has a large mass, it has a lot of matter in it. If something has a small mass, it has only a little matter in it. Since the baseball has more matter in it, the baseball's mass is greater than the tennis ball's mass.

Why is mass so important? Well, the mass of a planet determines the amount of gravity the planet has. Do you remember learning about gravity in lesson 1? Gravity is a force that planets use to pull on things. The sun's gravity pulls on the planets, keeping them in their orbits. Earth's gravity pulls on you, keeping you on the ground so that you don't fly off into space. Even when you jump as high as you can, you don't go very far before Earth's gravity pulls you back down to the ground.

Think about what would happen if God had created Earth with less mass. If Earth had less mass, it would have less gravity. This would mean it wouldn't pull on us as hard. We would be much lighter. Running a mile would be easy. We could jump onto the roof of our home without any effort. We could jump up into any tree and if we fell, it wouldn't hurt very much.

| Solar System Planet | How Many Planets to Be as Massive as the Sun |
|---|---|
| Mercury | 6,053,809 |
| Venus | 408,539 |
| Earth | 332,959 |
| Mars | 3,100,181 |
| Jupiter | 1,048 |
| Saturn | 3,499 |
| Uranus | 23,169 |
| Neptune | 19,418 |

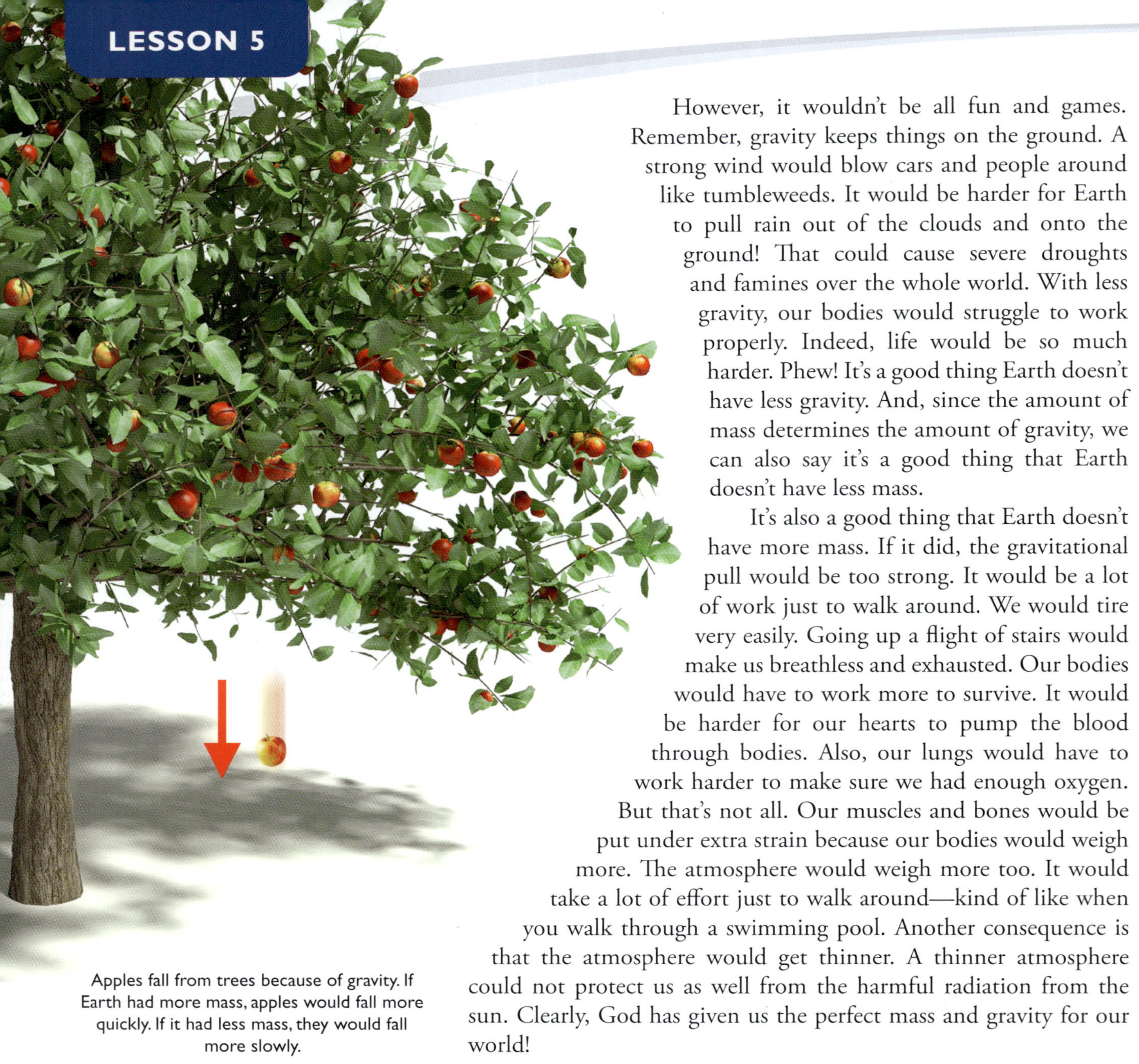

Apples fall from trees because of gravity. If Earth had more mass, apples would fall more quickly. If it had less mass, they would fall more slowly.

However, it wouldn't be all fun and games. Remember, gravity keeps things on the ground. A strong wind would blow cars and people around like tumbleweeds. It would be harder for Earth to pull rain out of the clouds and onto the ground! That could cause severe droughts and famines over the whole world. With less gravity, our bodies would struggle to work properly. Indeed, life would be so much harder. Phew! It's a good thing Earth doesn't have less gravity. And, since the amount of mass determines the amount of gravity, we can also say it's a good thing that Earth doesn't have less mass.

It's also a good thing that Earth doesn't have more mass. If it did, the gravitational pull would be too strong. It would be a lot of work just to walk around. We would tire very easily. Going up a flight of stairs would make us breathless and exhausted. Our bodies would have to work more to survive. It would be harder for our hearts to pump the blood through bodies. Also, our lungs would have to work harder to make sure we had enough oxygen. But that's not all. Our muscles and bones would be put under extra strain because our bodies would weigh more. The atmosphere would weigh more too. It would take a lot of effort just to walk around—kind of like when you walk through a swimming pool. Another consequence is that the atmosphere would get thinner. A thinner atmosphere could not protect us as well from the harmful radiation from the sun. Clearly, God has given us the perfect mass and gravity for our world!

**Can you explain in your own words all that you have learned about Earth so far? What is mass? How does it affect gravity?**

## Perfect Rotation

Another special feature of Earth is its **rotational period**. A rotational period is how long it takes a planet to make a complete turn. Since a planet's rotation turns night into day, a planet's rotational period is the length of one full day on that planet. Earth's rotational period is perfectly timed so that when we get tired, it gets dark outside. When we have slept enough, the sun comes up, making it light again. We have a 24-hour day, with about 12 hours of daylight and 12 hours of darkness. That's not the best thing about Earth's perfect rotational period, however!

Do you realize that how quickly or slowly a planet turns also affects the weather? If Earth rotated faster, it would have an effect on the winds; they would be stronger. On the other hand, if Earth rotated too slowly (giving us longer days and longer nights) the temperature difference between night and day would be extremely hard on the environment. You see, long nights are colder, and long days are hotter. A longer rotational period would result in colder nights and hotter days.

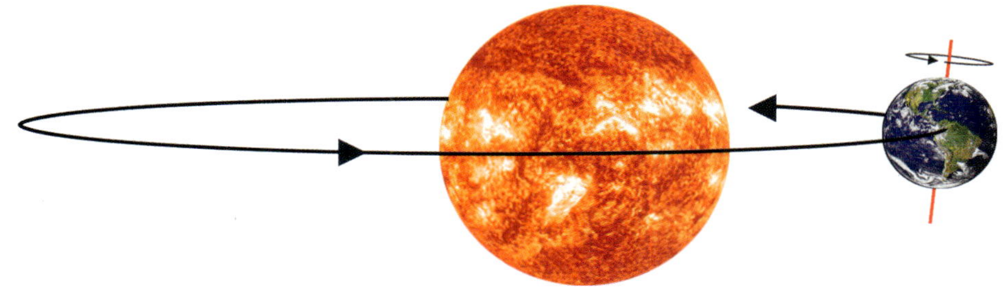

Earth rotates so that the sun rises in the east and sets in the west.

## think about this

The rotational period that God ordained for Earth keeps the temperature in balance to protect the people, animals, and plants that He created for His glory. The Bible says in Isaiah 45:18, "He is the God who formed the earth and made it, He established it and did not create it a waste place, but formed it to be inhabited." It isn't by random chance or accident that Earth has the perfect rotational period, the perfect mass, and the perfect position in space. God designed Earth as the perfect habitation for the entirety of His creation.

**Can you tell someone in your own words why the rotational period of a planet is important?**

# Perfect Atmosphere

God has also designed Earth with special chemicals in the atmosphere. We've already talked a little bit about what an atmosphere is. In Greek *atmos* means vapor and *sphaira* means sphere (ball). *Atmosphere*, then, means ball of vapor. That's why we call the mist and gases surrounding a planet the atmosphere of the planet. On Earth, we often call our atmosphere the air.

Even though you cannot see the atmosphere, it is definitely there! Take a deep breath. Do you realize you just sucked in lots and lots of **oxygen** (ox' uh jen) molecules? Oxygen is the most important gas in our atmosphere. Without oxygen, we would not be able to live. Astronauts take tanks of oxygen with them when they go out of our atmosphere. Their spacesuits pump this oxygen into their helmets so they can breathe it. If

An astronaut in space. Note the spacesuit (which has the oxygen tanks) and Earth below.

they did not have these oxygen tanks, astronauts would not be able to breathe in space. The spacesuits are like a small spaceship, creating an atmosphere similar to Earth's around the astronauts.

Have you ever been inside a **greenhouse**? A greenhouse is a house where gardeners grow plants. A greenhouse provides protection for fragile young plants from extreme temperatures. Most greenhouses have a roof that is milky white. This allows some of the sun's rays to shine on the plants, while protecting the plants from getting too hot. A greenhouse also keeps the plants from getting too cold. Our atmosphere is like a greenhouse that protects us from very high temperatures and very low temperatures.

This is another reason astronauts must wear spacesuits when they are in outer space. The spacesuits keep them warm. Once leaving our atmosphere, astronauts leave the comfort and warmth that God has provided us. Without a spacesuit, astronauts would freeze in outer space.

This picture, taken from a spacecraft orbiting at 200 miles above the surface, shows Earth's atmosphere as a thin blue band between the surface and the blackness of space.

Meteorite found near Fort Stockton, Texas, 1952.

Even though our atmosphere is over 75 miles thick above Earth (that's 396,000 feet!), unlike a greenhouse roof, Earth's atmosphere becomes thinner the higher it goes. The thinner the atmosphere gets, the more dangerous it is for people. This is why planes don't fly much higher than about 40,000 feet high and why only spaceships and astronauts can leave our atmosphere.

None of the other planets have the kind of atmosphere that we need to stay alive. Remember, Mercury doesn't have an atmosphere, and because of that, its surface freezes at night and burns hot during the day. Venus, on the other hand, has an atmosphere that would kill us if we breathed it. Also, Venus's atmosphere traps the planet's heat, keeping Venus extremely hot. As you learn about the other planets in our solar system, you will find that none of them have an atmosphere that would keep people alive. No other planet has the protection and security of an atmosphere like Earth. When God created Earth, He declared it *good*—and it is!

Imagine going outside one day to find that rocks were falling out of the sky. What if this happened every day? We wouldn't want to leave our homes, and if we had to, we would need steel umbrellas to protect ourselves! We would be afraid to look up at the sky, and our houses and buildings would have to be much stronger than they are now. Well, the fact is that rocks of every size—giant boulders, little pebbles and teeny tiny grains of sand—are constantly flying into Earth every day! Thankfully, the atmosphere that surrounds Earth burns most of these rocks into nothing but dust before they reach the ground. If God had not created our atmosphere just as He did, we would have to worry about getting hit by space rocks.

Have you ever seen a **shooting star**? Some people might call it a *falling* star. Well, it turns out that shooting (or falling) stars are not really stars at all! They are actually space rocks that burn up as they hit Earth's atmosphere. Occasionally, a piece of one of these space rocks survives the burning process and makes it to Earth's surface. If a piece does survive and reach the surface of Earth, it is called a **meteorite** (me' tee or eyet). Since oceans cover most of Earth, meteorites usually fall into the sea. It is rare for one to hit land. We will discuss meteorites and space rocks further in lesson 8.

**Try to put into your own words the things you have learned about Earth's atmosphere. How does it keep us warm? How does it keep us safe?**

## Perfect Tilt

What is the temperature usually like in the summer? What is it like in the winter? It feels warmer in the summer and colder in the winter, doesn't it? You probably already know that we have 4 seasons on Earth: winter, spring, summer, and fall. But, do you know why we have different seasons? You might think that maybe it's hot in the summer because we are closer to the sun in the summer. That is not true. Earth is actually a tiny bit farther away from the sun when it is summer in the United States! The real reason it's hotter in the summer is that the sun's light shines more directly on the northern hemisphere during the summer. It's the concentration and aim of the sun's light that makes it hot. In the other seasons of the year, the sun's light is not shining directly on us and does not warm our portion of Earth as well. Why does this happen? Why doesn't the sun's light shine directly on us all of the time? Well, Earth is permanently tilted in one direction. The North Pole is not straight up north, and the South Pole is not straight down south in relation to the sun. Our planet tilts. The parts of Earth that get direct sunlight change with each season, as Earth revolves in its orbit around the sun. If you have a globe, you will see that it is tilted. When you turn it, the countries close to the bottom of the globe come up toward the middle just a bit, and the ones near the top of the globe come down toward the middle just a bit.

# Activity 5.1a

## Understanding the Seasons

### You will need:

- Flashlight
- Piece of paper
- Globe

### You will do:

1. Take a flashlight into a dark room and lay it on a table.
2. Take a piece of paper and hold it right in front of the flashlight. When you do that, the paper gets direct light. In other words, the light is intense, concentrated in one spot.
3. Now tilt the paper so that the flashlight shines on it at an angle. Notice that when you tilt the paper, the light is not as intense. It still shines on the paper, but the light is more spread out and less concentrated. That is what happens in the seasons. In the winter, the sun is not shining straight down on us. When we tilt away from the sun's rays, the rays scatter and give us less heat, which results in cooler temperatures. The direct, concentrated rays that we get in the summer give us more of the sun's energy, resulting in higher temperatures.
4. Repeat the exercise with your globe. Shine the flashlight directly on the globe, and then tilt the flashlight so that the rays spread out. There is more intense heat when the sun's rays are straight and direct.

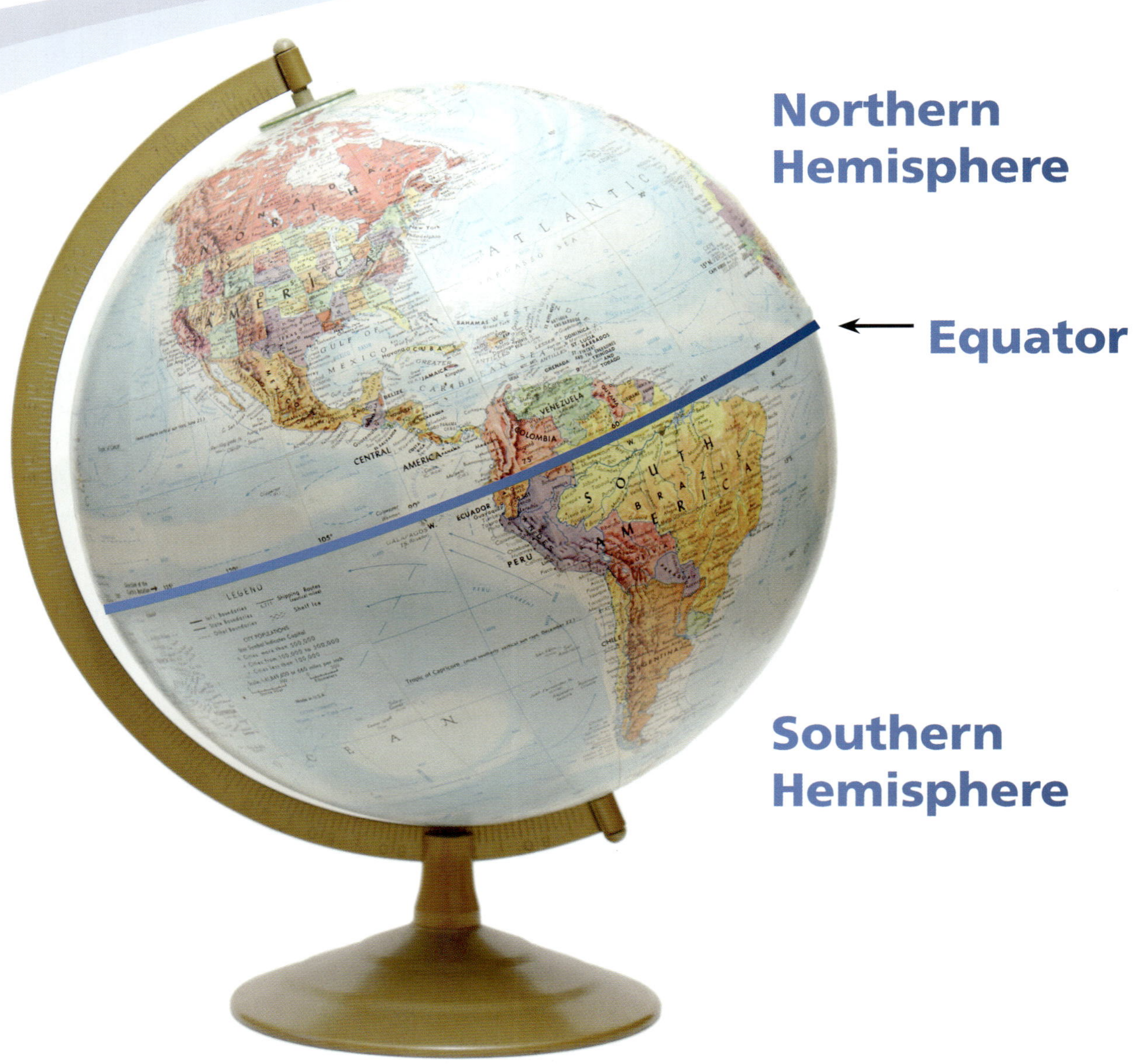


The **equator** (ih' kway tur) is the imaginary line that divides Earth in half. The sun shines almost directly on this middle line all the time. The top half of Earth is called the **Northern Hemisphere** (hem' uh sfear), and the bottom half is called the **Southern Hemisphere**. As Earth revolves around the sun, the hemisphere that points toward the sun gets direct sunlight, while the hemisphere pointing away from the sun gets less direct sunlight. In June, for example, the Northern Hemisphere tilts toward the sun. As a result, the sun's rays strike the Northern Hemisphere directly, and it is warm there. Since the Northern Hemisphere points towards the sun during this time, the Southern Hemisphere points away from the sun. How do you suppose it feels in the Southern Hemisphere during this time? That's right! The Southern Hemisphere experiences winter in June. In December, on the other hand, the Northern Hemisphere tilts away from the sun. This causes the sun's rays to strike the northern half of Earth less directly. This makes it cooler in the Northern Hemisphere, which is why the Northern Hemisphere experiences winter in December. Of course, at the same time, the Southern Hemisphere points toward the sun. So during December, it is summer in the Southern Hemisphere!

# Earth has seasons because its axis is tilted. Earth rotates on its axis as it orbits the sun, but the axis always points in the same direction.

## In December:

It's summer south of the equator, winter north of the equator. The sun shines directly on the Southern Hemisphere and indirectly on the Northern Hemisphere.

## In March:

It's fall south of the equator, spring north of the equator. The sun shines equally on the Southern and Northern Hemispheres.

## In June:

It's winter south of the equator, summer north of the equator. The sun shines directly on the Northern Hemisphere and indirectly on the Southern Hemisphere.

## In September:

It's spring south of the equator, fall north of the equator. The sun shines equally on the Southern and Northern Hemispheres.

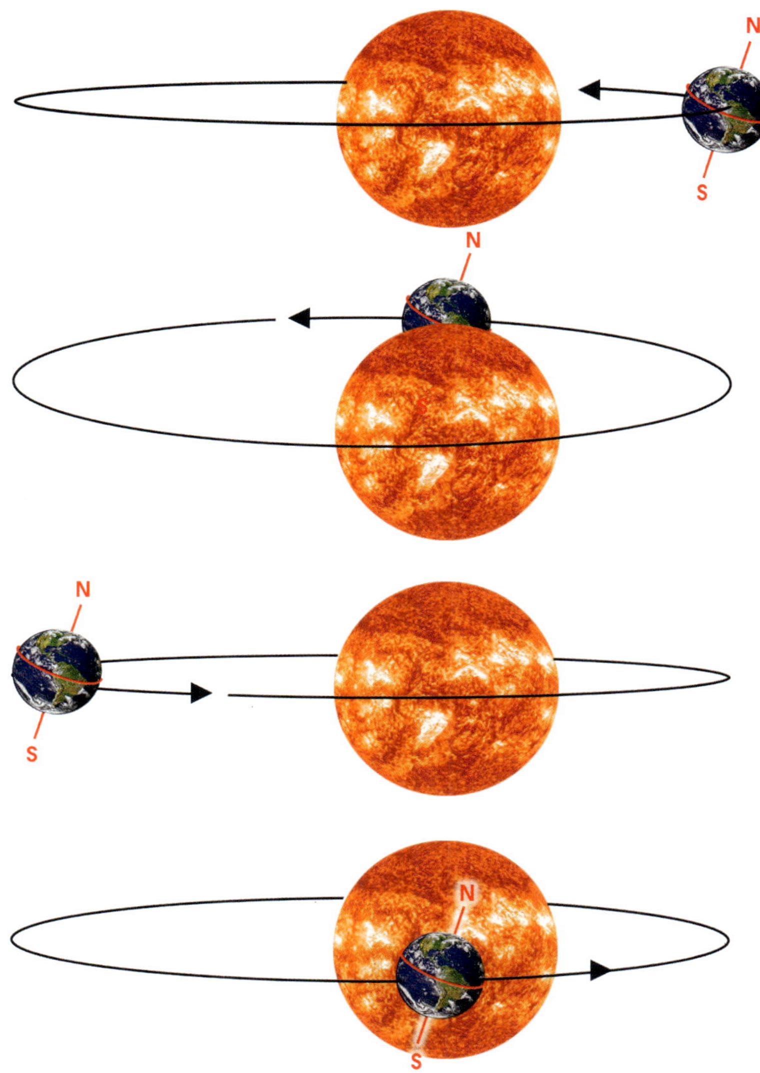


Remember that the sun shines more directly on the equator as compared to most of the other parts of Earth. This means that countries near the equator get the most direct sun rays. What do you think the temperature is like in those countries? They are nice places to go on vacation because they are warm all year. Find the countries close to the equator on your globe. Places that are close to the equator are warmer than those that are further away. Those who live in the southern United States, for example, are closer to the equator and get more direct sunlight, which means all of the seasons are warmer. In the southern United States, summer is very hot compared to the northern United States, and winter is reasonably warm compared to the northern United States.

# Activity 5.1b
## Understanding the Seasons

### You will need:
- Lamp
- Globe

### You will do:
1. Take the shade off a lamp and place it in the center of a dim room. The lamp is your sun.
2. Walk your globe around the lamp, as if it is Earth orbiting the sun. As you walk around the lamp, notice the 4 walls in your room. Each wall will represent a different season.
3. Start so that your globe is between the lamp and the center of one of the walls. We will say that this wall represents winter, so you need to hold your globe so that the Northern Hemisphere is pointing away from the lamp. Notice how the light from the lamp hits the Southern Hemisphere of the globe more directly than it does the Northern Hemisphere. This is why it is summer in the Southern Hemisphere while it is winter in the Northern Hemisphere.

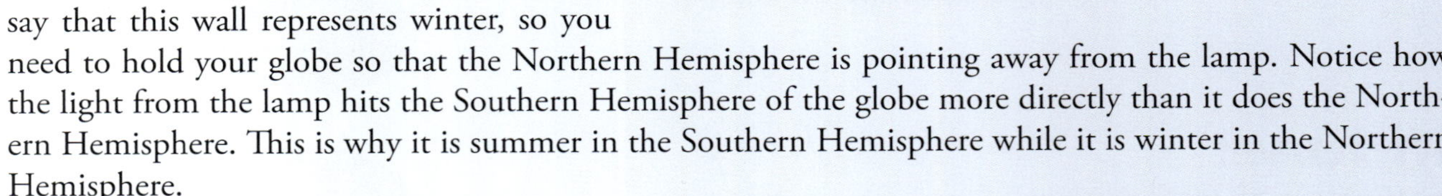

4. Now walk around to the lamp counterclockwise (in the opposite direction of the hands on a clock). Do not change the tilt of the globe. Just watch how the light hits the globe differently as you walk. When you reach the center of the next wall, you will see how light hits the Earth when it is spring in the Northern Hemisphere and fall in the Southern Hemisphere.

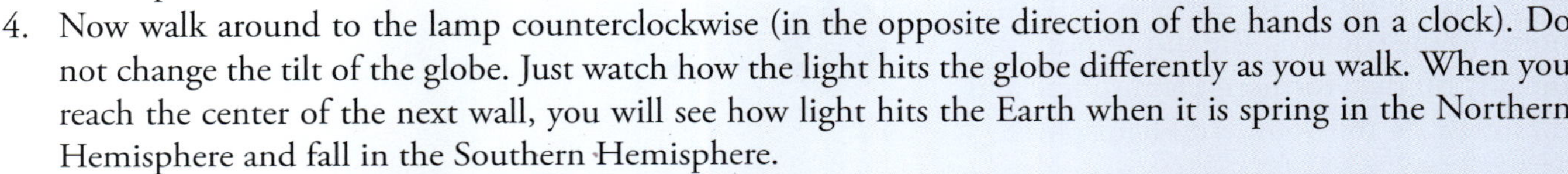

5. Continue to walk around the lamp counterclockwise, once again making sure that you don't change the tilt of the globe. When you reach the center of the third wall, you will see how light hits Earth when it is summer in the Northern Hemisphere and winter in the Southern Hemisphere. Notice how the lamp's light is hitting the Northern Hemisphere directly now. That's why it is summer in the Northern Hemisphere and winter in the Southern Hemisphere.
6. Continue walking around the lamp until you reach the center of the last wall. At that point, you will see how light hits the Earth when it is fall in the Northern Hemisphere and spring in the Southern Hemisphere.

The sun never shines directly on the countries near the top and bottom of the world. Even when the Northern Hemisphere is tilted toward the sun, for example, the North Pole is still too far from the equator to get direct sunlight. What do you think the temperature is like there? Even during the summer, it is still pretty cold (37 °F to 54 °F). At the very bottom of Earth is an icy continent called Antarctica. Like the North Pole, it never gets any direct sunlight, and is very cold. Even when the Southern Hemisphere is pointed toward the sun, temperatures average around 20 °F. When the Southern Hemisphere is pointed away from the sun, temperatures can get to be 128 °F below zero. During the winter, Antarctica is tilted so far away from the sun that it's dark all day and all night for months at a time. Then, when it's summer in Antarctica, it's tilted so that it always faces the sun and is light all day and all night. Can you believe that somewhere here on Earth the sun doesn't go down every night for months at a time? Of course, even though it is light all of the time in Antarctica during the summer, it is not very bright outside. Instead, the sunlight is just a dim glow. Can you explain why?

In summer, the days are longer than the nights. This is because the sun is shining down so directly that it takes longer for Earth to rotate enough for us to get out of the sun's light. Also, the sun stays higher in the sky during the summer because we are tilted toward it. In winter, on the other hand, the days are shorter than the nights. This is because we are tilted away from the sun, so it doesn't take long for Earth to rotate out of the sun's light. In winter, the sun also appears low in the sky.

Earth's tilt is very important. If Earth were not tilted, there would be no seasons. The northern United States would always be colder, while the southern United States would always be warmer. That would be a problem because we grow our food in the summer and allow the ground to rest in the winter. This allows the land to replenish its nutrients, which are like vitamins for the crops. Also, the cold winter kills the bugs that ruin crops. The colder the winter, the fewer bugs there are in the summer. More than just useful, the seasons make our world beautiful. Many flowers, like tulips, must have a cold winter in order to bloom in the spring. The seasons are part of what makes Earth a good place to live.

What is your favorite thing about the cold winter? What is your favorite thing about the warm summer? Without seasons, you could not enjoy those things. Aren't you glad God put Earth on a tilt and created seasons?

Despite what some people think, igloos made of snow are usually used as emergency shelters, not permanent homes.

**Explain how we have different seasons in your own words. Remember to include information about Antarctica and the North Pole in your explanation.**

# Perfect Land

When I was a kid, my siblings and I had a great big hole in our yard that we dug deeper and deeper every day. Our hole was so deep we could climb inside it. We planned to dig until we reached China! We didn't understand about the temperatures in the center of Earth. We didn't know that if we *could* dig a hole far down into the center of Earth, we would have to go through thousands of miles of molten rock before we ever hit China. The hole that we dug was very deep, but we didn't even get through the *first layer* of Earth.

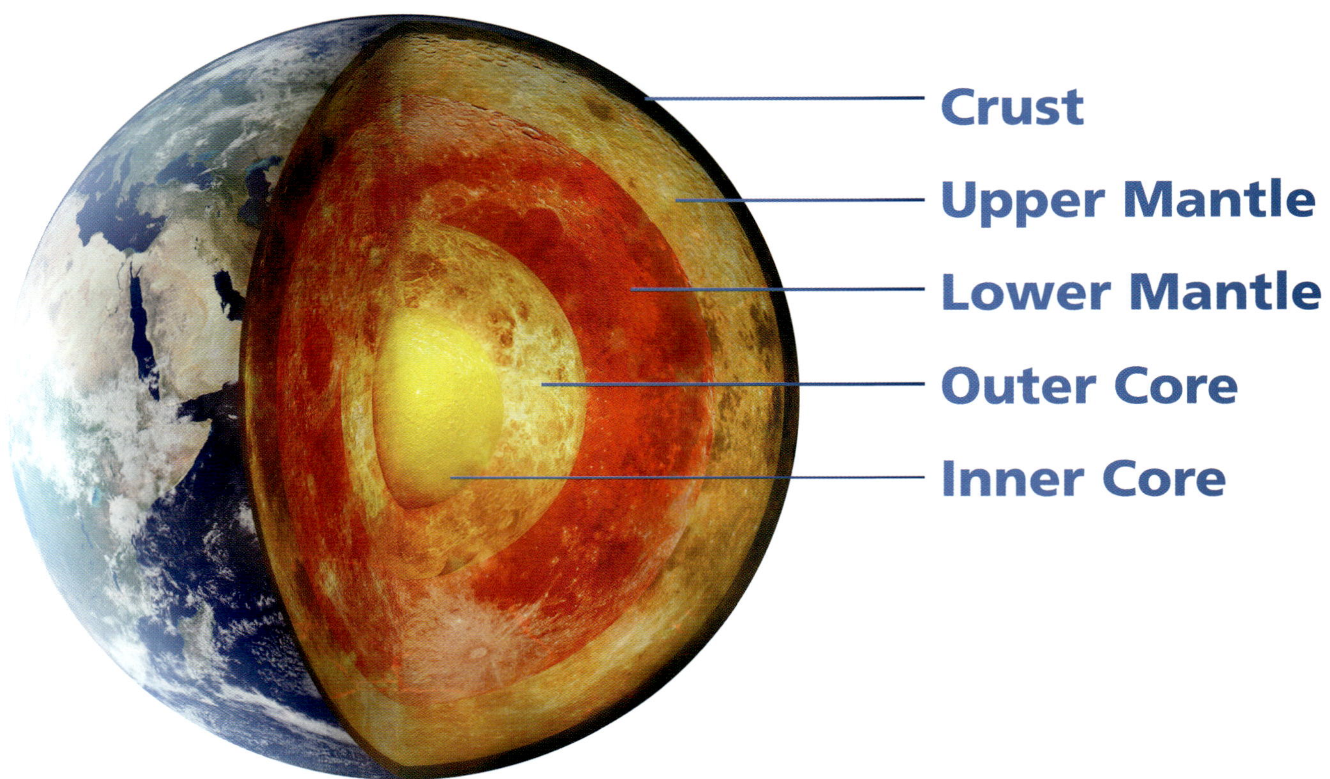


Our planet has many layers. The top layer is called Earth's **crust**, and it contains the oceans, dirt, rocks, and mountains. When you dig a hole in your backyard, you are digging in Earth's crust. Under the crust, there is a layer called the **mantle**. To reach the mantle, you would have to dig a hole roughly *20 miles deep*. The mantle is made of hot, semisolid rock, and it is thousands of miles thick. In fact, the mantle is the thickest portion of Earth. The farther down you go, the hotter and hotter the mantle gets. Magma is melted rock. There is a lot of magma in the mantle because it is so hot. Even below where you are sitting right now, there is hot magma. Don't worry; it is too far down for you to ever see it or feel it. In some places, however, holes in the crust allow magma to come up to Earth's surface from the mantle. These are volcanoes.

Below the mantle is the **core**. There is a hot section of melted metals like nickel and iron called the outer core. This important part of Earth forms Earth's magnetic field, which you will learn more about in the next section. At the very center of Earth are more nickel and iron, but here they are solid. This solid center is called the inner core of Earth.

**Can you name the layers of the Earth?**
**Can you tell someone at least one thing about each layer?**

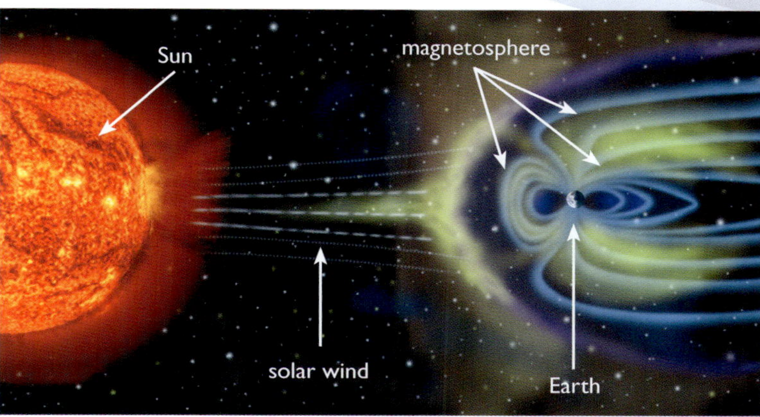


## Perfect Magnetosphere

Perfect magnet-o-what? Well, sphere means ball. Our **magnetosphere** (mag net' uh sfear) is like a big ball of magnetic power around Earth. It's an amazing, miraculous, mighty feature of our Earth and reminds us that God thought of everything! He didn't leave any little detail to chance.

Have you ever played with magnets? A magnet attracts certain metal objects. Have you ever noticed that when such an object gets close enough to the magnet, it jumps right onto the magnet? That's because the magnet has a magnetic field surrounding it, and the object is attracted to the magnet by that magnetic field. Can you believe the entire Earth has a magnetic field as well? This magnetic field is produced in the outer core of Earth, and it pulls certain harmful particles away from Earth. These harmful particles make up what scientists call the solar wind. God placed this special magnetic field around Earth to protect us from the dangerous particles coming from the sun. Our magnetosphere does an important and life-saving job.

Although the magnetosphere blocks most of the solar wind coming from the sun, some of the particles that make up the solar wind get trapped in the magnetic field, mostly around the North Pole and the South Pole. As these particles travel in Earth's magnetic field, they collide with gases in Earth's atmosphere. The energy of the collisions between these particles and the atmosphere's gases produces beautiful colors in the sky. These displays of color are called **auroras** (uh roar' uhs), and are easiest to see if you live in the northern or southern parts of the globe.

**Explain Earth's magnetic field in your own words.**

Auroras as seen from different parts of the Earth.

# Activity 5.2

## Make a Compass

A compass is an instrument that always points north. The reason it does this is because Earth has a powerful magnetic field that pulls the needle of the compass toward the magnetic North Pole. In this project, you are going to create your own working compass.

### You will need:

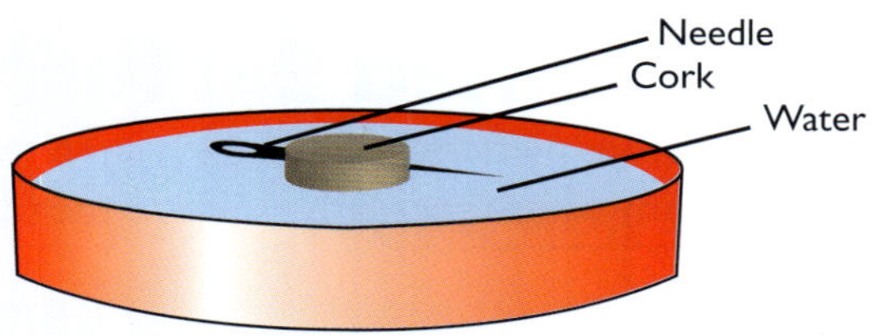


- Adult supervision
- Cork
- Permanent marker
- Lid from a yogurt or sour cream container (with a high lip)
- Sewing needle
- Magnet

### You will do:

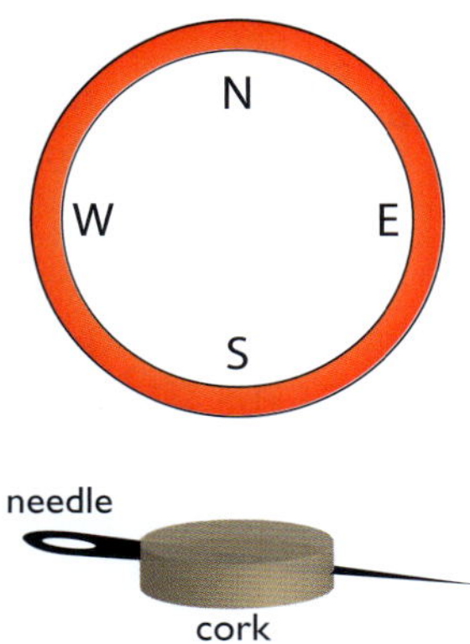


1. Use the marker to label each section of the lid with the first letter of each major direction: north, east, south and west. See the drawing to the right.
2. Run the magnet over the needle a few times, moving it <u>in the same direction each time it passes over the needle</u>. Do not run the magnet back and forth over the needle. Moving the magnet in the same direction magnetizes the needle a bit. In other words, it makes the needle a magnet.
3. Cut off a small, thin sliver from one end of the cork, and poke the needle through it, from one end of the circle to the other. See the drawing to the right.
4. Fill the lid with water.
5. Float the cork and needle in the lid so the floating needle lies roughly parallel to the surface of the water. You now have a compass.
6. Place your compass on a still surface and watch what happens.
7. Turn the compass, and notice that the needle continues pointing in the same direction, regardless of how you turn the compass. One end of the needle will always point north, and the other end will always point south.
8. Turn the compass so that the needle is pointing to the "N" that you drew in step 1.

### Discussion

Because you made the needle into a magnet, it is affected by Earth's magnetic field. As a result, one end will always point north, and the other end will always point south.

## *think about this*

Take a moment to think about how incredible our Earth is. Everything needed to sustain life on Earth did not happen by chance. The Bible tells us in Jeremiah 10:12, *It is He who made the earth by His power, Who established the world by His wisdom;And by His understanding He has stretched out the heavens.*

# Who Named Earth

While the other planets in our solar system are named after Roman and Greek gods and goddesses, Earth gets its name from what it actually is; *earth* means dirt or ground.

Astronomers use this symbol for Earth.

# Spacecraft for Studying Earth

There have been over 71 spacecraft missions sent into space to study Earth. The reason we send so many missions to space is to answer important questions about Earth. Some of the questions that these scientists want to find answers for are:

1. How is Earth changing?
2. How will Earth change in the future?
3. How can we prevent Earth changes?

One of these missions is named *AIM*. This mission is being used by NASA scientists to study the highest clouds in the atmosphere and what these clouds can tell us about Earth's climate.

You can see from the photograph that this spacecraft is not very large. Although it is not very big, it gives us valuable information.

NASA's *AIM* spacecraft being built.

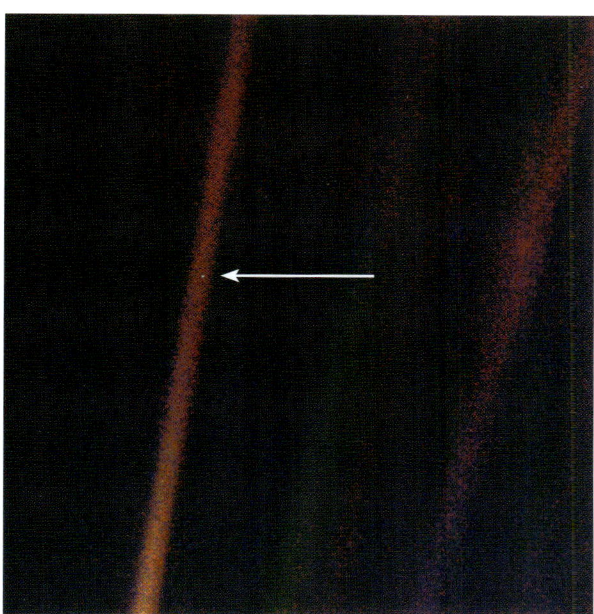

Pale Blue Dot (Earth) as captured by *Voyager 1*.

A larger spacecraft being used by NASA to study Earth was named *TERRA*. *TERRA* is an international mission carrying instruments from the United States, Japan, and Canada and is roughly the size of a small school bus. It studies Earth's water, land and atmosphere.

In 1977 NASA launched *Voyager 1* to explore our solar system. In 1990, when it was about 4 billion miles away, the spacecraft turned and took a series of photographs producing this image of our Earth. This is our home in the vastness of space!

# Activity 5.3

## Create an Advertisement for Earth

Today you are going to make an advertisement. If you do not know what an advertisement is, ask your parent or teacher to show you one in a newspaper or magazine. Pretend you have been asked to sell Earth to someone who does not know much about it. Make a colorful, full-page ad listing all the great features of Earth. Be sure to include all the things you have learned.

## What Do You Remember?

What are the 7 things that make Earth the only planet that can support life? Try to explain why those things help us to live on Earth. Why do we have different seasons? What are the 4 major layers of Earth? What was your favorite part of this lesson?

LESSON 6

# THE MOON

## *wisdom from above*

While you might be used to looking up at the night sky and admiring the Moon, did you know that God gave us the Moon for a purpose? The Bible tells us:

*"He made the moon for the seasons"*

Psalm 104:19

# The Moon

Isn't it wonderful to look up at the sky on a clear night? The Moon brightens up the night sky. Another word we use when talking about the Moon is **lunar** (loo' nur). The Latin word for Moon is *luna*, which is why we often use the word *lunar* to refer to the Moon.

Do you remember that the Moon is a natural satellite of Earth? The Moon revolves around Earth, like Earth revolves around the sun. Take a ball, hold it above your head, and move it slowly around your head. The ball orbits your head like the Moon orbits Earth.

The Moon looks like a big light up in the sky. But, actually, the Moon is a very dark satellite with no light coming from it at all. Why, then, is it lit up? Well, remember why we can see the planets shining like stars in the night sky. They reflect light that comes from the sun. Well, the Moon does the same thing. The sun is always shining on some part of the Moon, so it is always daytime somewhere on the Moon. The light we see coming from the Moon is the light of the sun reflecting off its surface.

Even though the Moon doesn't make its own light, it can be very bright in the night sky. When the Moon is big and round, the night is not very dark because the sun's light reflecting off the Moon provides some light by which we can see. When the Moon is a small sliver, we have darker nights. If you want to play hide-and-seek during the night, it is easier to hide when the Moon is either not visible in the sky or when it is just a small sliver.

**Tell someone all that you have learned about the Moon in your own words.**

# Activity 6.1
## Understanding How the Moon Reflects Light

### You will need:
- Lamp
- Compact disc

### You will do:
1. Take a compact disc (CD) into a room that has a lamp or a closet with a light bulb nearby.
2. Remove the lamp shade from the lamp, and turn on the light. Now move your CD in away that makes the light reflecting off of the CD onto your empty hand.

### Discussion
You are reflecting the light from the bulb (which is like the sun) off the CD (which is like the Moon) onto your hand (which is like Earth.) The CD is not producing light on its own. If you took the CD into a darkroom, no light would come from it. The only way the CD can shine light onto your hand is to reflect it from a lamp. In the same way, the moon reflects the light from the sun onto Earth.

## The Moon's Phases

Have you ever noticed that the Moon seems to change shape in the sky? Sometimes it's a big round ball, sometimes it's a semi-circle, and sometimes it's a sliver, or crescent. Also, one night during every month, the Moon seems to completely disappear.

A photograph of both Earth and the moon reflecting the sun's light.

Well, the Moon never really disappears, and it doesn't change shape. It's always the big round ball that God created it to be. The reason it looks like it has changed shape is because we can only see the part of the Moon that reflects light back to Earth. Why do you think only part of Earth and part of the Moon is lit up in the picture on the right? It is because the sun's rays are only shining on half of Earth and half of the Moon. The bright side of Earth is experiencing daytime, while the dark side of Earth is experiencing night. The people on the dark side of Earth in this picture would see a full Moon because the entire face of the Moon is reflecting sunlight onto Earth.

Do you remember in lesson 4 when we discussed the fact that if you look through a telescope at Venus, you can see Venus change shape from day to day? These are known as the phases of Venus. Well, the reason that the Moon seems to change shape in the night sky is that it has phases as well.

# Phases of the Moon as Seen from Earth

View is looking down at the Earth from the North Pole.

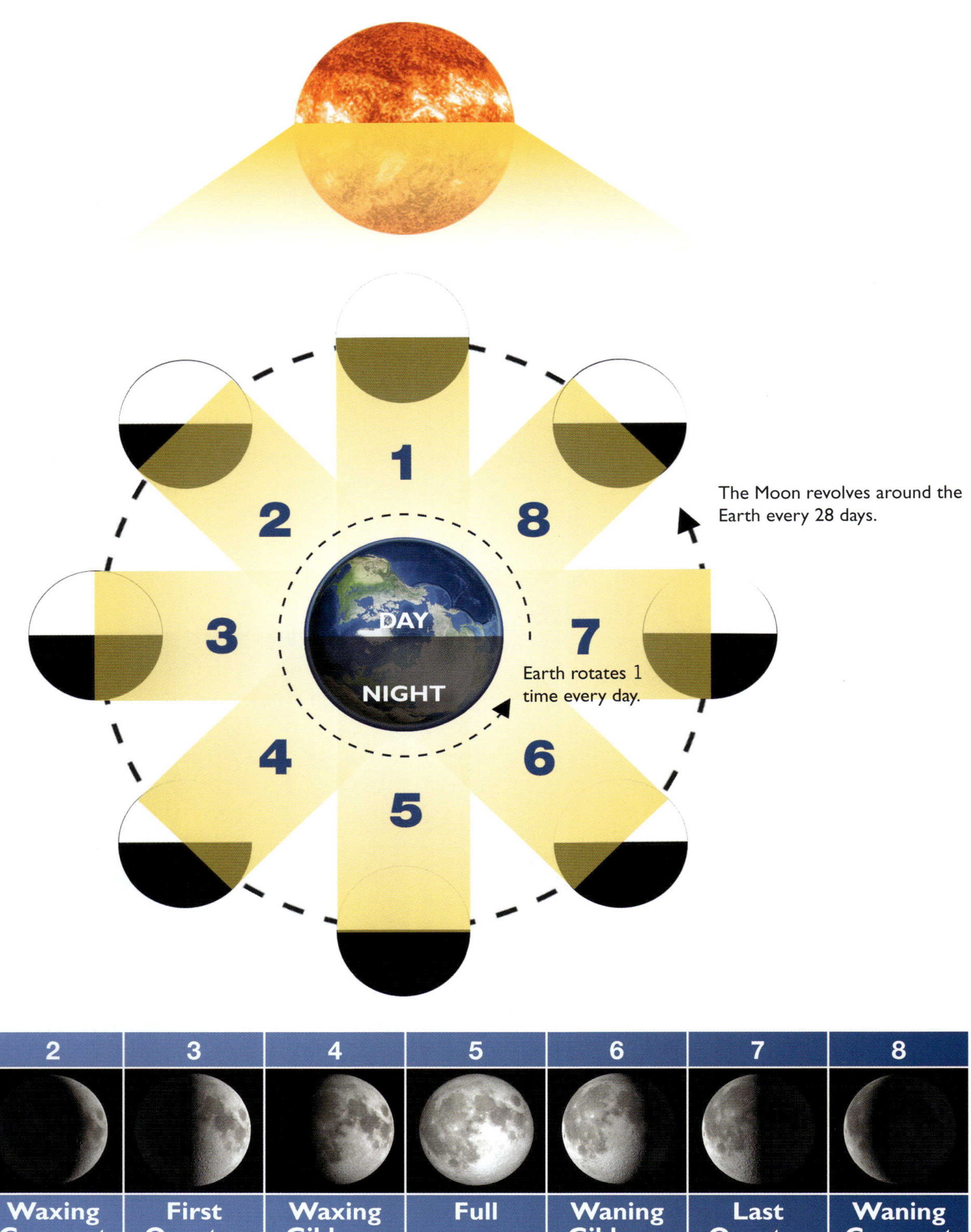


The best way to understand how the phases of the Moon work is to do an activity.

# Activity 6.2a
## Understanding the Phases of the Moon

### You will need:
- Lamp
- Lightly-colored ball (like a ping pong ball or white Styrofoam™)
- Stick
- Tape

### You will do:
1. Use the tape to attach your ball to your stick.
2. Put the lamp with its shade removed at one end of a darkened room. Sit at the other end of the room and hold the ball on the stick up in front of you so that it is between your face and the lamp and just slightly above your head. In this exercise, the lamp is the sun, the ball is the Moon, and your head is Earth.
3. Now keep your arm straight and slowly spin around in place so that you are constantly looking at the ball and so that the ball is traveling in a circle around where you are sitting. Do this very slowly so you can see how different sections of the ball are lit. As the ball travels in a circle around where you are sitting, you will see it go through the same phases that the Moon goes through. When the ball is between you and the lamp, you see only the dark side of the ball. When the Moon is between Earth and the sun, it is called a **new moon**. We can't see a new moon because it's totally dark. The night side of the Moon is the side facing Earth, and the day side of the Moon is reflecting light back to the sun.
4. Now look what happens as the ball moves so that it is no longer between the lamp and your face. As it moves, you first see that a small sliver of the ball reflects the lamp's light into your eyes. In the same way, when the Moon moves so that it is not directly between Earth and the sun, we begin to see a little sliver of the Moon. This little sliver is called the **crescent moon**.
5. As you continue to spin around, you will get to the point where you can see a semicircle of light reflecting off the ball. When the Moon reaches that point, it looks like a semicircle in the sky. That is called a **quarter moon**. It's not called a half moon, even though you see it as a half-circle. This is because the word quarter refers to the fact that at this point, the Moon is 1/4 (or a quarter) of the way through its orbit around Earth.
6. As you continue to spin around, you will see that the portion of the ball that reflects light gets larger and larger. When the Moon starts getting bigger and bigger after the quarter Moon, it's called a **gibbous moon**.
7. As you continue to turn, you will eventually reach the point where your head is in between the lamp and the ball. Because the ball is slightly above your head, however, your head does not block the light from the lamp, and you see the entire side of the ball shining with the lamp's light. When the Earth is between the Moon and the sun, we see the entire side shining down on us, so we call it a **full moon**.
8. As you continue to spin around, you will see the pattern reverse. The bright part of the ball will get smaller and smaller. Eventually, you will see only half a circle, then a sliver, then a dark ball again.

### Discussion
The Moon goes through exactly the same phases that your ball went through, and it takes about 28 days. That means it takes 28 days for the Moon to circle Earth. After that, the whole process starts over again.

# Activity 6.2b

## Record the Phases of the Moon

Every night (and sometimes during the day), look at the Moon outside and draw its shape on a calendar. Write down the name of the phase. You can refer to the previous illustration to determine what the shape is called.

PLEASE NOTE: Because of Earth's rotation, your location on Earth faces the Moon during specific times. As a result, the Moon rises and sets, just like the sun. Sometimes, the Moon cannot be seen after sunset because it sets before the sun. During those times, you can see it before sunrise. You can also see it during the day. The course website, **www.apologia.com/bookextras**, has a link to a site that will tell you when the Moon will rise and set in your area. Use this resource to plan when to look for the Moon. Try to find the Moon in the daytime during this activity as well as at night, so that you will understand that the Moon rises and sets, just like the sun.

## *think about this*

Our Moon revolves around the Earth in an egg-shaped orbit called an ellipse. The closest point to Earth is called *perigee* and the farthest point is called *apogee*. When a full moon is at perigee, it is about 14% larger than usual. We call this a *supermoon*.

**Explain to someone in your own words why the Moon reflects light. Be sure to explain how the reflection creates the phases of the Moon.**

# Lunar Eclipse

Every once in a while, as it is making its way around Earth, the Moon lines up perfectly with the sun and Earth, as shown in the illustration below. When this happens, you could draw a straight line from the center of the sun, through Earth, and to the Moon. This causes a **lunar eclipse**. Do you remember what a solar eclipse is? It's when the Moon gets in between the sun and Earth, blocking out the sun. A lunar eclipse is when Earth gets right in between the Moon and the sun, blocking the sun's light from the Moon.

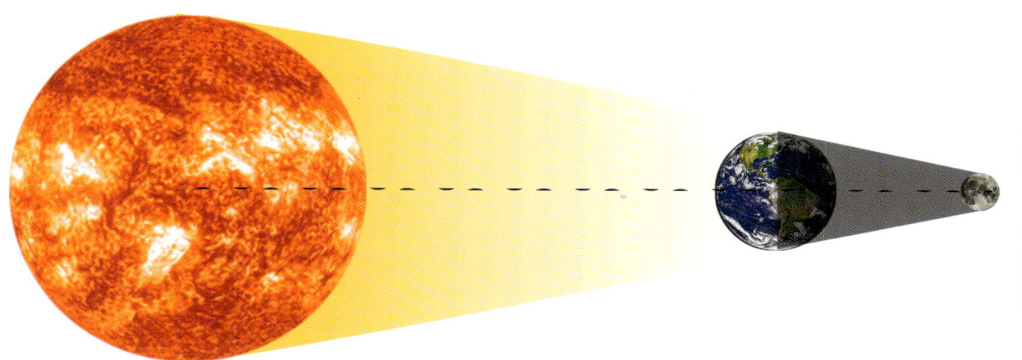

**Lunar Eclipse Dates**

January 31, 2018
January 21, 2019
May 26, 2021
May 16, 2022
November 8, 2022
March 14, 2025

You should never look directly at a *solar* eclipse. However, you can stare right at a lunar eclipse, and it's a beautiful thing to see! When a lunar eclipse occurs, the shadow of Earth crosses over the Moon. If you watch a lunar eclipse as it happens, you will see a circular shadow slowly pass over the Moon. It will start at one side of the Moon and gradually cover more and more of the Moon. In a partial lunar eclipse, the shadow covers only part of the Moon. In a total lunar eclipse, the shadow covers the entire Moon. The portion of the Moon covered in the shadow will be a beautiful reddish-copper color. This happens because the sun's light must travel through Earth's atmosphere before it shines on the Moon. Do you remember what happens when the sun's white light travels through Earth's atmosphere? It bounces off of the gases in the atmosphere. Since blue light tends to bounce off these gases more than other colors of light, the blue light does not make it through the atmosphere. As a result, it does not reach the Moon. Instead, the red, yellow, and orange light from the sun ends up striking the Moon, giving it the beautiful color shown in the picture.

A total lunar eclipse occurring. You can see Earth's curvature on the moon's surface as it blocks out the sun.

The moon during a total eclipse.

Do you understand why it is safe to look at a lunar eclipse, even though it is quite dangerous to look at a solar eclipse? It's because the Moon does not produce any light. The only light that we see coming from the Moon is the sunlight that it reflects. Because of that, the amount of light that comes from the Moon is very small compared to the amount of light that comes from the sun. As a result, looking at a lunar eclipse is very safe, and you should try to see one yourself. Look at the dates of upcoming lunar eclipses in the blue box. You can learn more about how to see these upcoming lunar eclipses by visiting **www.apologia.com/bookextras**.

**Take a moment to explain how a lunar eclipse happens. Be sure to explain why is it safe to look directly at a lunar eclipse.**

# Lunar Atmosphere

A photograph of Astronaut Buzz Aldrin walking on the Moon. Notice that the sky is black when the sun is shining because the Moon has no atmosphere.

You might hear someone say that a restaurant "has no atmosphere." This means that the restaurant is kind of boring and not very cozy. Sometimes, you'll hear people say, "That place has a lot of atmosphere!" When people say that, they mean the place is warm and friendly. Sayings like these are just expressions, but they do contain a bit of truth. Earth's atmosphere does make Earth a warm and friendly place to live. Remember how our atmosphere is one important thing that protects us and gives us air to breathe? The Moon has no atmosphere. So, the Moon would not be a very warm or friendly place to live. We would definitely say, "This place has no atmosphere!"

If you went to the Moon, everything would be so different from Earth. Without an atmosphere, there would be no protection from the sun's rays. You would need an extra heavy spacesuit so you wouldn't get sunburned. The spacesuit would also carry the oxygen you would need in order to breathe. Because of this, you could only stay outside of your spaceship until your oxygen level got low. You would then have to head back for a refill so that you could explore some more.

Do you remember why Earth's sky is blue? The blue rays of the sun bounce off the gases in our atmosphere, making it look like blue light is coming from the entire sky. As a result, the sky looks blue. Since the Moon has no atmosphere, its sky looks black, even during the day. You should remember from lesson 3 that Mercury's sky is the same. It looks black even during the day because Mercury also has no atmosphere.

The maria on the surface of the Moon.

Do you remember how our atmosphere protects us from rocks that fall from outer space? Space rocks also fall from outer space onto the Moon. Without an atmosphere, all those flying rocks crash right into the Moon. Some have been so enormous that they have left huge dents on the surface, which we call craters. The Moon is scarred with thousands of these craters.

The Moon rotates very slowly. It takes just as long for the Moon to make one rotation as it does for the Moon to make one orbit around Earth. Because of this, the same side of the Moon is always facing Earth. Whenever you look up at the Moon, then, you always see the same side. Compared to the other side of the Moon, the side that faces Earth has very few craters. Instead, it is covered with smooth, flat plains called **maria** (mar' ee uh). They appear as dark patches on the Moon's surface.

**Explain to someone else some facts you know about the Moon's atmosphere.**

# The Moon's Gravity

A spacesuit seems like it would be very heavy doesn't it? Interestingly, your spacesuit would not feel very heavy on the Moon. That's because there is a lot less gravity on the Moon compared to the gravity on Earth. Why? Do you remember what determines a planet's gravity? It is the mass of the planet. Since the mass of the Moon is a *lot* less than the mass of Earth, the Moon's gravity is much weaker than Earth's gravity.

Even though the Moon's gravity is weaker than Earth's gravity, it is still there. Because of that, if you were to drop a ball on the Moon, it would fall to the ground. However, it would not fall nearly as quickly as it would on Earth. If you jumped up as high as you could on the Moon, you would still end up falling back to the ground. However, you would go up a lot higher than you could here on Earth, and you would not fall back to the ground nearly as quickly as you would on Earth. You could jump up and do a somersault in midair and not hurt yourself at all, even if you didn't land on your feet. That's because the Moon's gravity would not pull you down as hard as Earth's gravity does.

You also would not weigh very much on the Moon. In fact you would weigh about 1/6 as much as you do on Earth. Look at the chart to the right to see if you can figure out how much you would weigh on the Moon.

| Your Weight on Earth | Your Weight on the Moon |
|---|---|
| 20 pounds | 3.3 pounds |
| 30 pounds | 5.0 pounds |
| 40 pounds | 6.6 pounds |
| 50 pounds | 8.3 pounds |
| 60 pounds | 10.0 pounds |
| 70 pounds | 11.6 pounds |
| 80 pounds | 13.3 pounds |
| 90 pounds | 14.9 pounds |
| 100 pounds | 16.6 pounds |
| 150 pounds | 24.9 pounds |

Even though the Moon may seem far away, the Moon's gravity affects us here on Earth! If you have ever been to the beach, you might remember walking a long way just to get to the water. If you visited that same beach at another time, you might have noticed that the water was closer. Even when you stand in the same place on the beach, sometimes the water is near you and sometimes the water is far away. This is because the ocean has **tides**. When the water is close, we say that the ocean is at **high tide**. When it is farther out, we say that the ocean is at **low tide**.

Believe it or not, the Moon's gravity is what causes the tides. Do you know what *bulge* means? If you put your stuffed animals in your shirt, it would bulge. To *bulge* is to curve outward or protrude. As the Moon's gravity pulls on Earth's oceans, they bulge as shown in the diagram below.

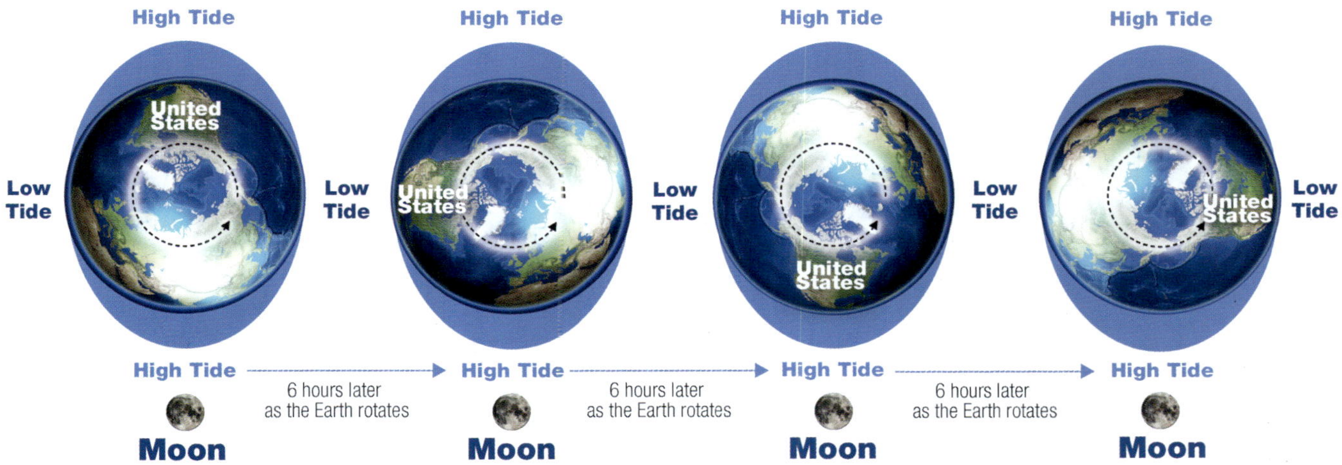


Looking down on Earth's North Pole.

Do you remember what rotate and revolve mean? Since Earth rotates faster than the Moon revolves around it, each point on Earth passes through high tide (where the ocean bulges) twice a day and low tide twice a day. Look at the picture of the tides. The Moon pulls on the oceans causing them to bulge while Earth rotates through the bulged areas.

If you have visited the beach, you have probably observed that the ocean has waves. Although the tides can cause certain types of waves, the waves that we see on the shore are not caused by the tides. They are usually caused by wind. Don't confuse tides with waves. Tides determine how close the ocean water is to you when you are on a beach. At high tide, the water is much closer to you than at low tide. Waves cause the ocean water to heave back and forth. They do not really affect how close the ocean is to you.

If the Moon were no longer in the sky, our oceans would not have tides. God gave us a Moon that has enough gravity to pull on the ocean's water. As the tide comes in and goes out, it cleanses the shoreline. The motion caused by the tides also refreshes the water. Still water can get stagnant, and the tides keep the ocean from getting dirty. God designed Earth with a Moon to keep the oceans fresh and clean.

## think about this

In addition to looking really pretty in our night sky and creating tides, the Moon also stabilizes Earth's wobble! Without the Moon, the Earth's rotational axis from east to west could change over a long period time.

*Oh, the depth of the riches both of the wisdom and knowledge of God! How unsearchable are His judgments and unfathomable His ways!* Romans 11:33

# Activity 6.3
## Make a Telescope

This is a great exercise to learn about telescopes and how they work. If you want a telescope that will really be able to see far into the heavens, you will need to purchase one from a store. However, it is fun to create a working telescope that you can use to see the Moon (or any other distant object) better. This activity will give you an introduction to how telescopes are made.

**(Continued on next page.)**

## You will need:

- Adult supervision
- 2 magnifying glasses (One should be stronger than the other.)
- Construction paper
- Tape
- Scissors
- Tape measure
- Paper with writing or an image on it

## You will do:

1. Place the paper on a table or the floor.
2. Hold the stronger magnifying glass a good distance between you and the paper.
3. Look at the paper through the lens. The image will look blurry.
4. Place the weaker magnifying glass between your eye and the first magnifying glass.
5. Move the weaker magnifying glass forward or backward until the print or image comes into sharp focus. The print or image should appear larger and upside down.
6. Keep the magnifying glasses in place while you have someone measure and record the distance between the two magnifying glasses.
7. Make two paper tubes using your construction paper and the measurement you determined for their length. The first one will fit around your strongest magnifying lens. The second should fit around your weaker magnifying lens and be slightly smaller in diameter so that it can slide into the first tube.

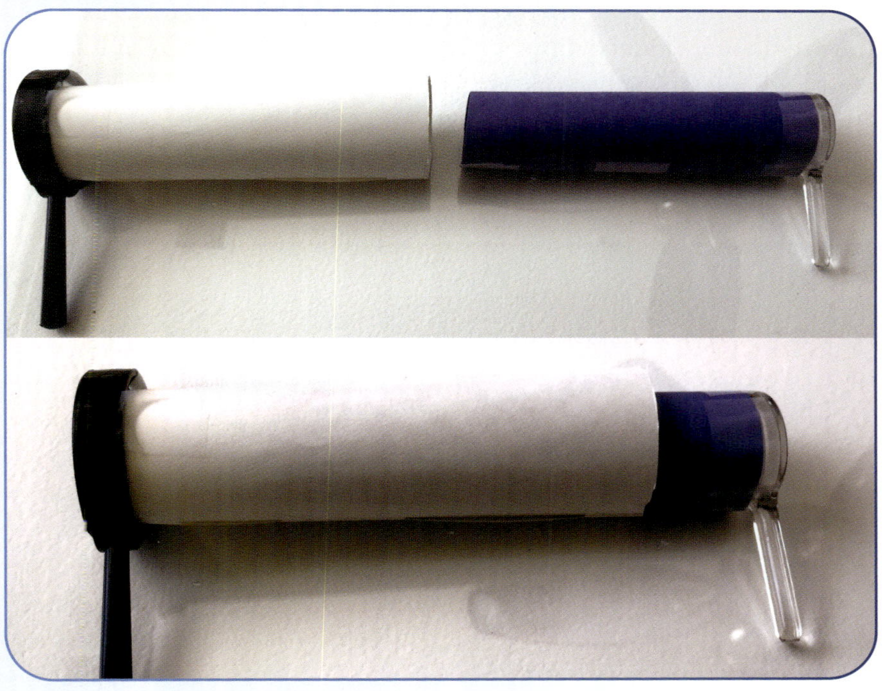

8. Secure the tubes to the magnifying lenses with your tape.
9. Look through the smaller magnification lens at something outside. **Never look directly at the sun!** If things are not in focus, you may have to slide your lenses apart.
10. Please note that just like the writing on the paper, the images you see through your telescope will be upside down. Despite the fact that they are upside down, the images should appear to be closer than they actually are because your telescope is magnifying them for you.
11. Try looking at the Moon with your telescope. Can you see any details that you were not able to see with just your eyes?

**(Continued on next page.)**

**NOTE: NEVER, NEVER LOOK AT THE SUN OR A BRIGHT OBJECT WITH THIS (OR ANY OTHER) TELESCOPE. It will severely hurt your eyes! You could even go blind! You must have special equipment if you want to look at the sun through a telescope.**

### Discussion

Your telescope works because the magnifying glasses bend the light that comes through them. The first magnifying glass (the one farthest from your eye) gathers the light from the object you are looking at and flips it upside down. The second magnifying glass (the one closest to your eye) takes the image and makes it bigger (magnifying it) for your eye. If you ever want to look at a distant object but do not have a telescope, you can just use two magnifying glasses to bring it into better view.

## Who Named the Moon

Ancient Romans called the Moon *Luna*, and part of that has stuck. Remember, you just learned about a lunar eclipse and the lunar atmosphere. Like Earth, however, we call our Moon what it is. It's the same with our sun. You will learn in later lessons that other moons and stars have proper names.

Astronomers use this symbol for the Moon.

## Spacecraft to the Moon

In the early 1960s, President John F. Kennedy asked NASA to send a man to the Moon and to bring him safely back to Earth. To achieve this goal, NASA started a program called Project Apollo. The Apollo program needed a mighty rocket so NASA developed the *Saturn V* rocket (*V* means five in Roman numerals). A picture of this rocket is shown to the left. It was longer than a football field!

In addition to the rocket, NASA developed two spaceships. One was called the Command and Service Module (CSM) and the other was named the Lunar Module (LM). Both were attached to the top of *Saturn V*.

The Command Module carried three astronauts. These three astronauts circled around the Moon and then came back to Earth. When the spaceship got back to Earth, it landed in the ocean.

The Service Module was where the rockets, the fuel tanks, and life support systems were located. It also had the antenna which was used for long-distance communications to Earth. It did not return to Earth. The Command and Service Module is shown in the picture on the left.

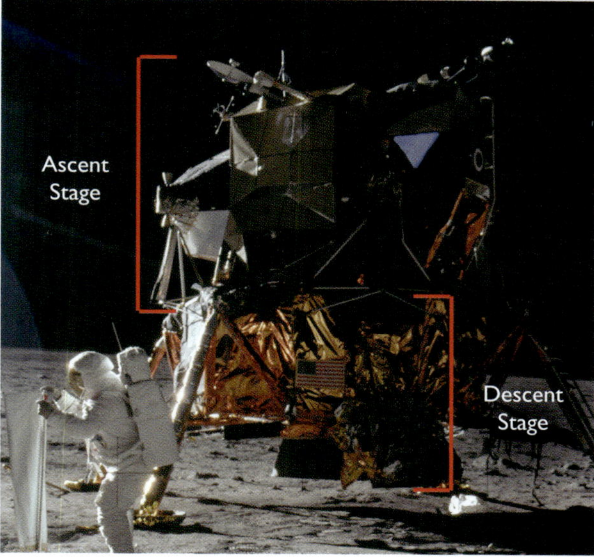


The Lunar Module could land two astronauts on the Moon. It could also leave the Moon and go back up to meet with the Command Module. However, the Lunar Module was not designed to return to Earth. Once in space, it stayed there.

The Lunar Module had 2 parts; the ascent stage and the descent stage. Each stage had its own engine. The descent stage had equipment to study the surface of the Moon. The crew cabin where the astronauts stayed was in the ascent stage.

The Moon is the only place in our whole solar system, other than Earth that people have visited. The first man who walked on the Moon was an American named Neil Armstrong. He flew to the Moon on the *Apollo 11* and first stepped onto the Moon on July 20, 1969. Do you see the footprint in the picture to the right? That is an astronaut's footprint in the dirt on the Moon. Even though this footprint was left on the Moon quite some time ago, it is most likely still there. When you leave footprints in the dirt in your backyard, wind will eventually blow the dirt around, filling in your footprints. Rain might also smooth out the dirt, once again destroying your footprints. On the Moon, however, there is no wind or rain. Because of that, the footprints do not get filled in. If people visit the Moon again sometime in the future, they will probably see the same footprints left by astronauts in 1969.

A photograph of an astronaut riding the lunar rover on the moon.

In all, 9 Apollo missions went to the Moon. Can you take a guess how many men have walked on the Moon? 12! Some of these men not only walked on the Moon, but they rode a special car that they took with them. It was called the **lunar rover**, and it allowed them to explore more of the Moon.

## think about this

Even though American astronauts have placed 6 American flags on the Moon, the United States does not own it. An international law written in 1967 prevents any single nation from owning planets, stars or any other natural objects in space.

There is a plaque (sign) on our Moon that reads:
*Here men from the planet Earth first set foot upon the Moon July 1969, A.D. We came in peace for all mankind.*

An artist's drawing of the *LRO* orbiting the Moon.

Do you want to walk on the Moon? If you become an astronaut, you might walk on the Moon, or even a planet like Mars!

Can you believe that 58 spacecraft have gone to the Moon so far? These spacecraft were *unmanned*; they had no people on board. Why have so many spacecraft been sent to the Moon? Some measure the Moon's gravitational field. Some take pictures of the surface. They've also been sent to look for water.

Right now, NASA has two spacecraft orbiting the Moon. These spacecraft are named *LRO* and *LADEE*. *LRO* is making a complete map of the Moon's surface and *LADEE* is studying the surface of the Moon.

**Take a moment to tell someone about lunar exploration. You can use the pictures to help you remember.**

## What Do You Remember?

Can you explain why the Moon has phases? What is a lunar eclipse? What is the atmosphere like on the Moon? What is the color of the Moon's sky during its daytime? How does the Moon affect Earth's oceans? Why is that helpful to Earth? Why are there footprints likely visible on the Moon? What was your favorite part of this lesson?

LESSON 7

# MARS

## wisdom from above

We have now passed Earth and are moving on to explore Mars! Mercury had no atmosphere; Venus had too much atmosphere. Earth had perfect atmosphere. Mars is yet another wonder of creation.

*How precious also are Your thoughts to me, O God! How vast is the sum of them!*

Psalm 139:17

# Mars

Mars is a terrestrial planet, like Earth. It has landforms like mountains, valleys, and volcanoes. One Martian (mar' shin) volcano called **Olympus** (oh lim' pus) **Mons** (mahns) is the largest volcano in our whole solar system. In fact, it is 3 times higher than Mount Everest on Earth! It is so big that astronomers on Earth can see it through a telescope. We have never seen a Martian volcano erupt, but there is evidence that some of the volcanoes are active, which means that they *could* erupt. Scientists are not even sure what would come out of the volcanoes on Mars if they did erupt.

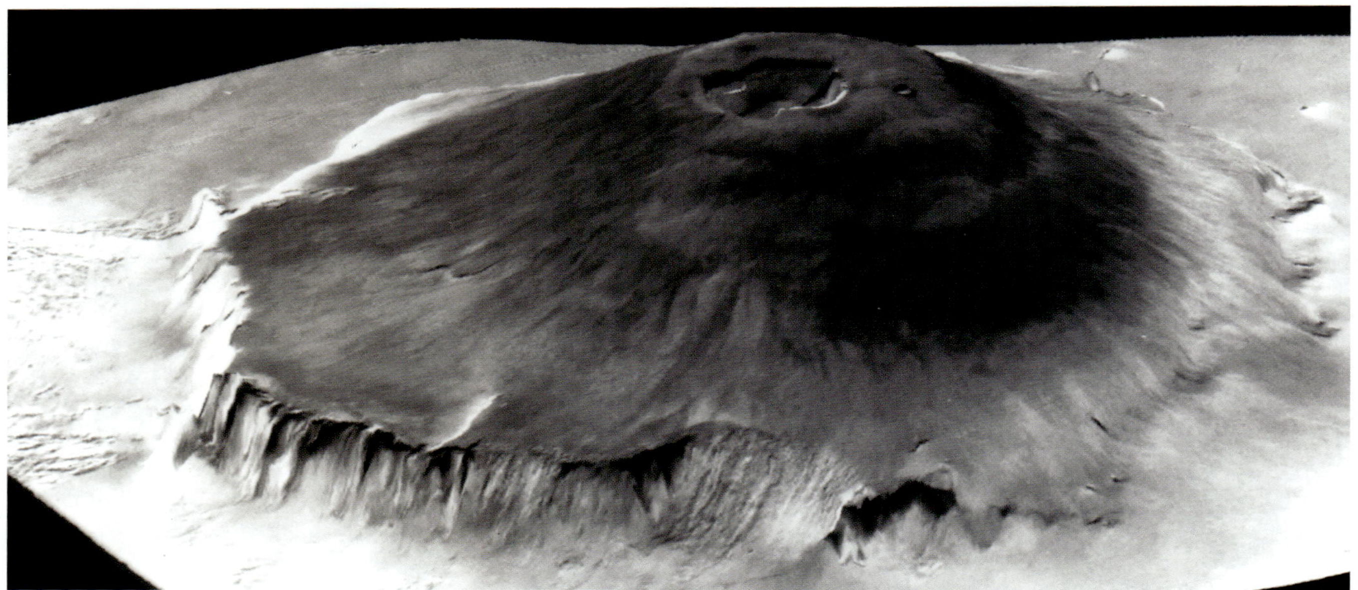

Olympus Mons. It is about the size of the state of Arizona.

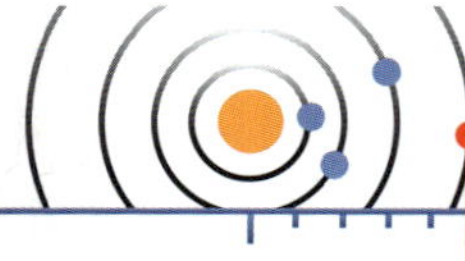

Mars is 147 million miles away from the sun.

# Activity 7.1

## Build Olympus Mons

The biggest mountain peak in the entire solar system is on Mars. It is a volcano called Olympus Mons. Below is a photo looking down on the volcano from above. In this project, we are going to build a volcano that really erupts!

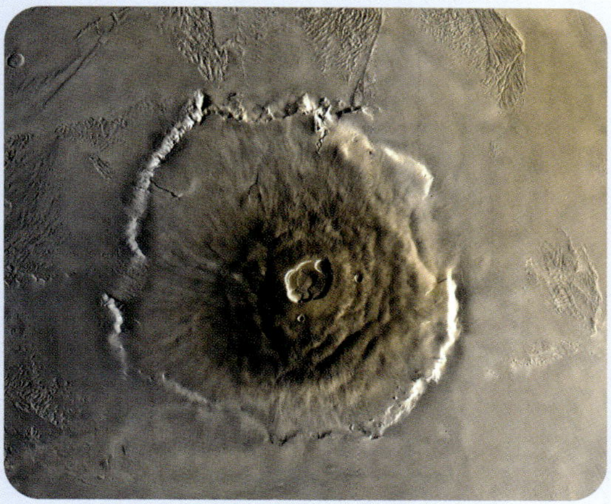

### You will need:

- Salt dough recipe ingredients (see below)
- Small bowl
- Alka-Seltzer® tablet
- Rocks (optional)
- Red and yellow food coloring

### You will do:

1. Make the salt dough according to the recipe to the right.
2. Shape your dough into a volcano that will fit inside your bowl.
3. If you have rocks, use them to surround your volcano, so that it looks like the surface of a rocky planet (see photo).
4. When you have a nice mountain, let it dry for the rest of the day. When it is dry, it will look like Olympus Mons, the biggest volcano in our solar system!
5. The next day, carefully break up your Alka-Seltzer® tablet into small pieces and place them inside your volcano.
6. Add a few drops of red and yellow food coloring (or whatever color you think Mars's volcanic eruptions would be) to a small amount of water.
7. Pour the colored water into your volcano and stand back!

**Salt Dough Recipe**

In a small bowl, mix together:

½ cup flour
2 T salt
½ tsp. cooking oil
2 T water
2 drops red food coloring

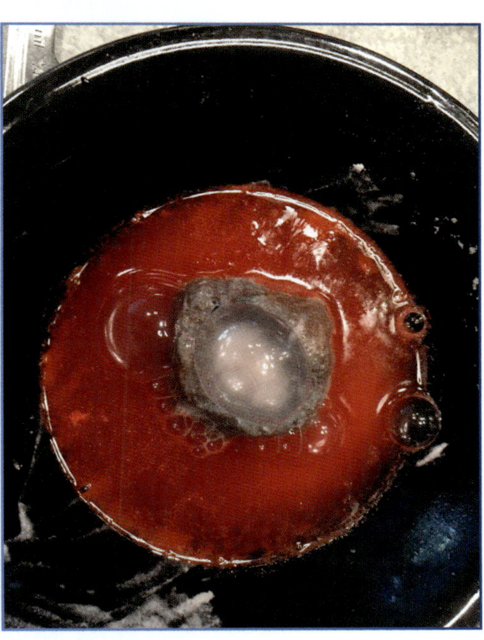

### Discussion

You can experiment with different amounts of materials and colors to see which eruption you like best.

## Martian Gravity

Mars is bigger than our Moon, but smaller than Earth. It's small for a planet, and its mass is significantly less than the mass of Earth. Because its mass is small, the gravity on Mars is not very strong. Think again about what it would be like to live with less gravity. What kinds of things would you have to be careful about if you lived on Mars? With less gravity, a baseball could probably go a lot further than when thrown on Earth.

How much would you weigh on Mars? Check the chart that compares weights on Earth and Mars. Moving to Mars might not be so bad if you weighed less. It wouldn't hurt as bad if you fell off your bike. It would also be easier to get around if you weighed less. You wouldn't get nearly as tired at the end of the day. Maybe moving to Mars isn't such a bad idea after all!

| Your Weight on Earth | Your Weight on Mars | Your Weight on Earth | Your Weight on Mars |
|---|---|---|---|
| 20 pounds | 7.6 pounds | 70 pounds | 26.5 pounds |
| 30 pounds | 11.4 pounds | 80 pounds | 30.3 pounds |
| 40 pounds | 15.2 pounds | 90 pounds | 34.1 pounds |
| 50 pounds | 19.0 pounds | 100 pounds | 37.9 pounds |
| 60 pounds | 22.8 pounds | 150 pounds | 56.9 pounds |

**Take a moment to review with someone. What can you tell them about the surface of Mars? What can you tell them about its gravity?**

## Martian Atmosphere

Mars as photographed by the *Spirit* rover. Notice the sky color.

Mars has a very thin atmosphere with very little oxygen. That is why we would need to take our own air if we went to Mars. Flying through the atmosphere on Mars are millions of little dust particles, just like in our atmosphere. These dust particles reflect the yellow, red, and orange light from the sun. Guess what color the sky is on Mars. It's a yellow/brown color, like butterscotch candy!

A Martian sunset as photographed by the *Pathfinder* rover.

Of course, just like on Earth, the sky changes colors when the sun rises and sets. The dust particles in the Martian atmosphere make the Martian sunset a blue color, as shown in in the photograph to the left.

Do you remember that our atmosphere keeps meteors from hitting our planet? Mars's atmosphere is too thin to offer much protection from flying space rocks.

All of the craters on Mars suggest that, at one time, Mars was pelted with giant space rocks. When this happened, the force of the collisions probably sent many pieces of Mars flying out of its atmosphere and into space. Some of those pieces of Mars actually landed on Earth! We know this because we have actually found rocks on Earth that have chemicals in them which are very similar to the chemicals found in Martian rocks. These rocks look like they have fallen at high speeds through our atmosphere, so it is reasonable to assume that they actually came from Mars!

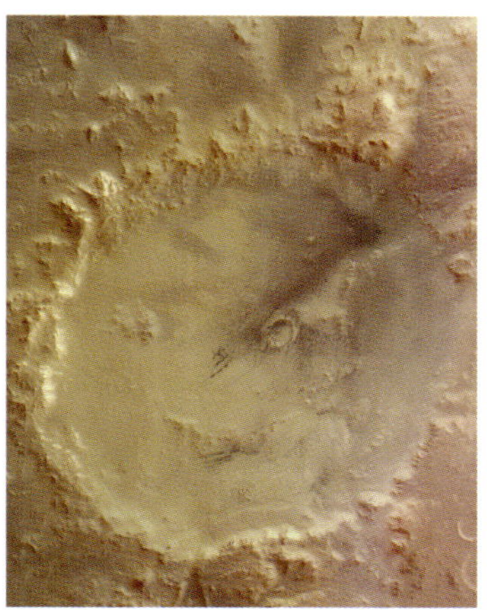
Can you tell why this large Martian crater is called the *happy face crater?*

A meteorite stone which scientists believe came from Mars.

It is likely that there are also pieces of Earth on Mars. If a giant meteorite hit Earth, the force would be strong enough to send pieces of Earth up into our atmosphere and out into space. Since there are also craters on Earth, we know this has happened before. Now remember, there are not many craters on Earth because Earth's atmosphere protects us from most space rocks. However, some space rocks are so large that the atmosphere cannot destroy them, and they land on Earth, making a huge crater. Creation scientists believe that if anyone ever finds signs of life on Mars, it will not be Martian life they find, but Earth life that made it to Mars! After all, if a piece of Earth left our atmosphere, it would take with it many cells and bacteria, which are living things.

A crater in the Arizona desert.

## God's Gift of Magnetism

Earth's magnetic field—the magnetosphere—is about 3,000 times stronger than the magnetic field on Mars. As we learned in lesson 5, this magnetic field shields us from the solar wind coming from the sun. Near Earth, the solar winds are trapped in a giant magnetic bubble and moved around Earth back out into space. You can see how this works by examining the picture of the solar winds around Earth and around Mars. As you know, without our magnetic field, life on Earth could not exist. Deadly radiation from the solar wind would strip away our atmosphere.

Early Mars is thought to have had a magnetic field strong enough to have allowed Mars to have a thicker atmosphere. This thicker atmosphere may have been able to keep the planet warm enough to allow liquid water to be on the surface. But from information sent from NASA spacecraft looking at the magnetic field lines preserved in Martian rock, scientists have discovered that Mars's magnetic field suddenly disappeared at some time in the past. Without a strong magnetic field to protect it, Mars's atmosphere is exposed directly to the solar wind which over time has helped thin the atmosphere to its current thickness.

### *think about this*

The Bible assures us that God made Earth for us to live on:

*For thus says the Lord, who created the heavens (He is the God who formed the earth and made it, He established it and did not create it a waste place, but formed it to be inhabited), "I am the Lord, and there is none else.*
Isaiah 45:18

**What do you remember about Mars so far? Explain in your own words the things you have learned. Be sure to talk about the atmosphere on Mars and the magnetosphere surrounding Mars.**

## Moons

If you were standing on Mars at night, you would see not one moon up in the sky, but 2! Mars is the only planet in our solar system that has 2 moons which astronomers have named **Phobos** (foh' bohs) and **Deimos** (dee' mohs). However, they don't look or act like anything God would have designed to be a moon. First, they are tiny and potato-shaped. Also, they don't orbit like normal moons. Phobos is getting closer and closer to

Mars every day. At some point, it might crash into Mars, although at the speed it is moving toward mars, it would take a long time (about 40 million years) for this to happen. For these reasons, astronomers believe that Phobos and Deimos were rocks floating free in outer space before getting caught in Mars's gravitational pull and becoming natural satellites.

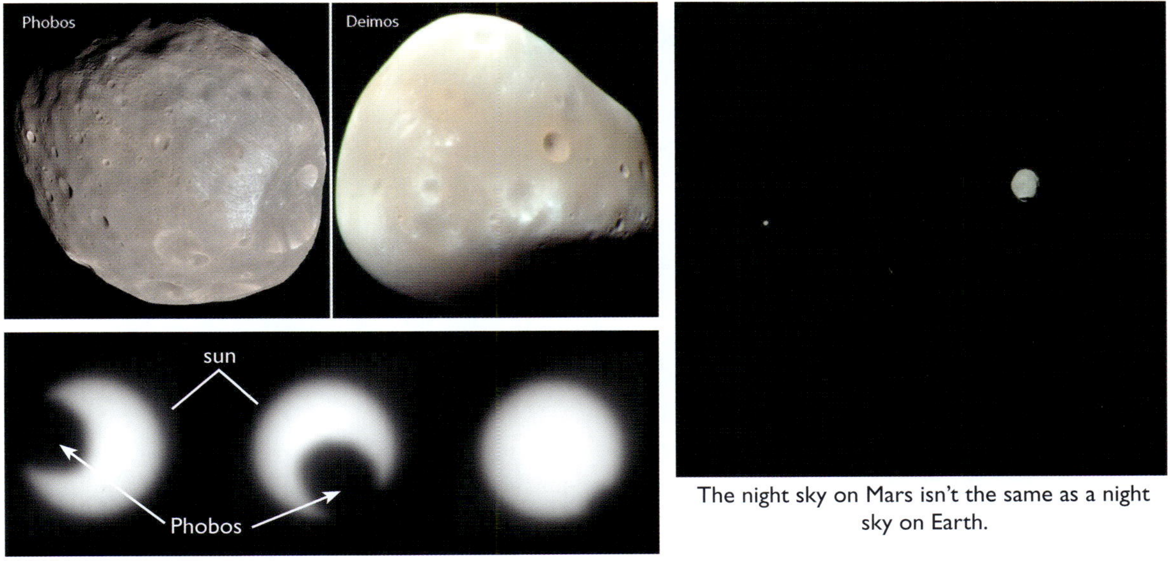


Phobos crossing between the sun and Mars.

The night sky on Mars isn't the same as a night sky on Earth.

Do you remember what a solar eclipse is? It's when the Moon blocks out the sun. Well, on Mars, the moons cannot completely block out the sun because they are too small. However, they can partially block out the sun. While on Mars, the *Opportunity* rover took these photographs of the moon Phobos crossing in front of the sun. You can see that the moon is too small to block out the entire sun.

## Martian Orbit

It takes almost 2 years for Mars to travel all the way around the sun. One year on Mars is 687 Earth days. If people eventually move to Mars, will they count their birthday by Earth days or Martian days? What about babies born on Mars? Will there be a different calendar for Martian colonists? If they decided to count time by Martian years, on your 10th birthday, you would almost be 20 Earth years old. If you lived on Mars, you could drive a car when you were only 8 1/2 Martian years old!

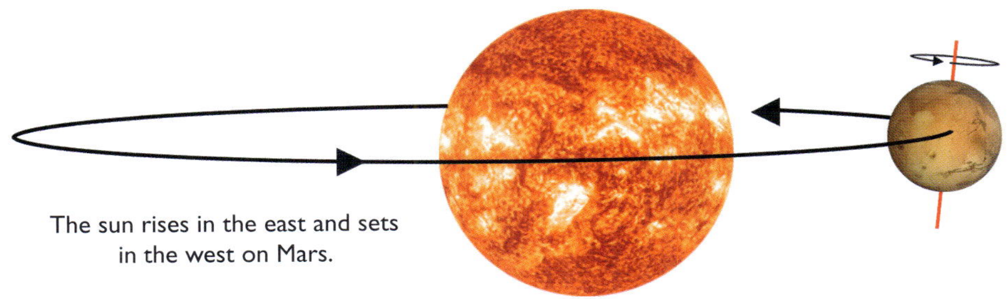

The sun rises in the east and sets in the west on Mars.

**Take a moment to review what you know about the 2 moons on Mars. Also talk about a martian orbit.**

# Martian Rotation

If you moved to Mars, the only thing that would not be a huge adjustment would be the number of hours in a day. Mars rotates in 24 hours and 39 minutes. Earth rotates in 24 hours. So, a day on Mars is just a bit longer than a day on Earth. That would make the whole experience of moving to Mars a little better. Moving to a planet that had days and days of sunlight followed by days and days of darkness would be incredibly difficult. Since Mars has days that are similar to Earth, Mars would be an easier planet to live on than Mercury (where the day is 59 Earth days long) or Venus (where the day is 243 Earth days long).

Mars rotates on a tilt, just as Earth does, so it has 4 different seasons; however, those seasons aren't like those we enjoy on Earth. Winter on Mars is freezing cold; spring is cold; summer is cold at the poles; and fall is cold. Mars is just too far away from the sun to get much energy from the sun's light, and its atmosphere is just too thin to keep the heat inside. The average temperature on Mars is usually about 81 °F below zero. The average temperature on Earth is 57 °F above zero.

Summers on Mars can be a lot warmer. But that doesn't mean you'd enjoy them. Imagine if when you were lying on the ground, it was a nice, comfortable 70 °F. Now, picture standing up and the temperature around your head was so cold your breath came out in steam and your spit froze in the air. That's what summers are like on Mars. At the equator it might be a lovely 70 °F during the day. But that's only close to the ground. Upon standing, you would find the temperature so cold around your head that you would need a stocking cap because at the height of a person's head, the temperature would be only 32 °F! Why is that? It has to do with the atmosphere. It's just too thin to keep the temperature the same at your feet and your head.

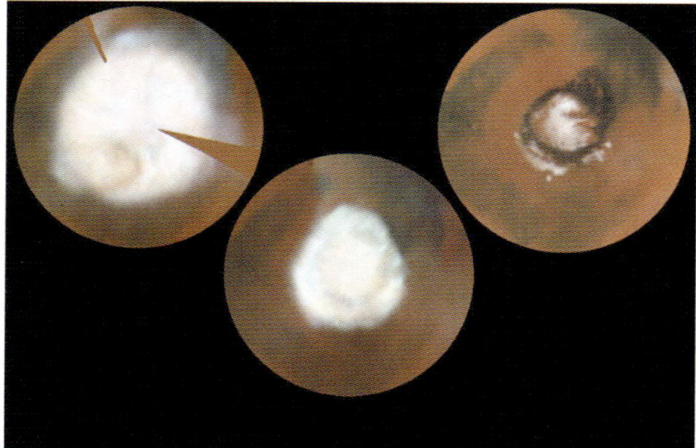

Three images of Mars's north pole. The white is ice. Most of the ice is not made of water. It is dry ice (solid carbon dioxide). The fact that the amount of dry ice changes demonstrates that Mars has seasons.

Temperature extremes at the Martian equator during the summer (in degrees Fahrenheit).

Mars can get super cold. In fact, in the winter, the very top and very bottom of Mars (the poles) can get as cold as 225 °F below zero. Mars's weather is a lot like the weather in Antarctica (you know…the place where nobody permanently lives because it is far too cold).

Have you ever seen **dry ice**? Dry ice looks a lot like regular ice, but it is much colder. It is so cold that you would hurt your hands if you tried to hold onto it for very long. Dry ice is actually frozen carbon (kar' bun) dioxide (dye ox' eyed), a gas that you and I breathe out every day. Carbon dioxide has to get very cold to freeze, so when it is frozen, it is much colder than normal ice that you make from water. It is so cold on Mars that even carbon dioxide freezes in the winter, creating a lot of dry ice.

The surface of Mars as photographed by NASA's *Viking Lander 2* in 1979. It shows a thin coating of water ice on the rocks and soil.

Scientists think that most of the water on Mars is frozen. Some of it is frozen in **polar icecaps**, which are big slabs of ice at the north and south poles of the planet. If you look at the picture on the opposite page, the ice that remains on the north pole during the Martian summer is mostly frozen water. The summer gets warm enough to get rid of the dry ice, but it does not get warm enough to melt the ice that is made of water. Scientists think that some of the water on Mars is permanently frozen in the ground. Because the water is permanently frozen in the ground, they call it **permafrost** (pur' muh frost). That sounds like "permanent frost," doesn't it? Well, that's exactly what scientists think it is! Recent discoveries show that liquid water is present in small amounts near the equator during certain times of a Martian year.

Mars is freezing, but it's closer to Earth's temperature than any other planet in the solar system. This is the reason Mars is the only planet in our solar system that we could consider sending humans to. It is also the reason that some scientists dream of building a habitat, or artificial ecosystem, that will protect a community where people can live, work, and grow food. Inside the artificial ecosystem, the temperature would be kept warm.

## Moving to Mars

Have you ever moved to a new home? It's difficult to get used to a new neighborhood and city. What if one day your parents told you to pack up your stuff because you were moving to the fourth planet from the sun? It sounds farfetched, but that is what many scientists hope will happen one day. Because Mars is similar to Earth in some ways, scientists have dreamed about building a community on Mars where people can live.

Although this might sound like a neat idea, there are a lot of problems with it. God did not give Mars all the incredible features that make Earth such a wonderful place to live. Because of this, you would have to take things from Earth that you need for survival, including water, air, heat, food, shelter, and clothing. In fact, you would have to bring just about everything because there is very little on Mars to support life! You would have to live in a special enclosed habitat designed by scientists, called an artificial ecosystem. This habitat would seal in the oxygen, allow the sun's light to come in, and provide enough water to drink as well as grow plants. What do you think such an ecosystem might look like?

The painting on the previous page shows an ecosystem that NASA designed. NASA hopes to send people to Mars at some point, but it costs a great deal of money to send people to other planets. It may be many years before any person is sent to Mars. Would you like to take a trip to Mars? How would you like to move there?

Would you like to live on Mars for a few years? What would you miss about Earth? Would you miss the blue sky? Would you miss the warm breezy summer days? Would you miss swimming in the ocean? If you were to build a protected community on Mars, would you build an imitation ocean for people to swim in? Would you build it around one of Mars's mountains, planting trees and grass on the mountain and providing fake snow in the winter, so people could go mountain climbing and skiing? If people asked you what nationality you were, would you say, "I'm a Martian," or, "I'm an Earthling living on Mars."

# Activity 7.2

## Designing a Mars Community

Imagine and design a community where people could live on Mars. What would you need to build an ecosystem to protect people and provide the right temperatures to grow food? Draw a picture of it. You can use the Internet or encyclopedias to see what an ecosystem, such as the Biosphere 2 (see photo below) in Arizona, look like.

### You will need:

- Adult supervision
- Imagination
- Materials you choose

### You will do:

1. Use your imagination and different materials you find to build houses and buildings.
2. Create mountains.
3. Add items that you would need to survive on Mars.
4. Have fun displaying your habitat and explaining it to others.

## Mars Surface

Mars has always been called the *Red Planet*. Whether looking through a telescope, or without any visual aid equipment, Mars looks red. When I say "naked eye," I mean that you are just using your eyes (no telescope, binoculars, or other equipment) to look at it. Robots that have landed on Mars have studied its red dirt, and found that it has an element called iron in it. Iron rusts, and that is what makes Mars look red. Mars is, in fact, a rusty planet!

## Liquid Water on Mars?

We know that there is liquid water on Mars. However, unlike Earth, which has lakes, rivers, and oceans, the liquid water on Mars barely dampens the soil. NASA has sent 4 rovers to the surface of Mars. Their names are *Sojourner*, *Spirit*, *Opportunity*, and *Curiosity*. All of them have helped scientists learn a lot of things. Unfortunately, only 2 of the rovers (*Opportunity* and *Curiosity*) are still working and sending back information about Mars.

What's the big deal? Why worry about whether or not there is liquid water on Mars? Well, liquid water is essential for life, and some people want to believe that there was (and perhaps still is) life on Mars. If they cannot find any liquid water on Mars, or if they cannot find evidence that there *used to be* liquid water on Mars, it will be very hard for them to continue to believe that life once existed on Mars. Of course, just because liquid water has been found on Mars, that does not mean life did exist or does exist on Mars. It only means that it is possible for some form of life to have existed or possibly still exist on Mars.

A photo of rocks on Mars. The surface texture indicates that the rocks might have been in liquid water.

How can we tell if liquid water exists on Mars? Well, some rock formations are usually made in the presence of water. Also, there are certain chemicals that tend to form in liquid water. *Spirit* and *Opportunity* have found both of these things on Mars. The picture above is of a rock formation that *Opportunity* found on Mars. On Earth, you see this kind of texture in rock if the rock has been soaking in salty water. The *Spirit* rover also found traces of the mineral *jarosite*. On Earth, this mineral forms in liquid water.

Once again, it is important to realize because scientists have now found evidence of liquid water on the surface of Mars, that does not mean that life exists there. Liquid water is necessary for life, but so are many, many other things. You learned in lesson 5 that Earth has been perfectly designed for life. Mars does not have most of the design features that Earth has, so it is hard to understand how life could have or does exist on Mars. You also need to remember that since space rocks have hit Earth and thrown parts of Earth into space, it is possible that there are rocks from Earth on Mars. If that is the case, there may, indeed, be some kind of life on Mars (like bacteria), but that life might have actually come from Earth.

**Tell someone about the difficulties humans would have to overcome if they ever moved to Mars.**

Astronomers use this symbol for Mars.

## Finding Mars in the Sky

Mars is fairly easy to locate in the night sky. It is brighter than most stars, and it has a distinctly orange/red color. However, Mars moves around quite a bit in the sky. If you want to find its location at any given time, visit **www.apologia.com/bookextras**. On that website, you will find links that will help you find Mars.

## Who Named Mars

Because of its red color, Mars is named for the Roman god of war.

## Spacecraft to Mars

More than 40 spacecraft have visited Mars. These spacecraft were unmanned which means they did not have any people onboard. Some of these spacecraft orbit around Mars, while others have landed on the Red Planet.

Four of the spacecraft have landed on the surface. They have wheels so that they can move around. In fact, they're like little cars and are similar to toy remote control cars. They are called Mars rovers. *Sojourner* was the first rover to land on the Martian surface and arrived in 1997. This little rover sent back more than 550 pictures. It also gave us information about the Martian wind and weather. The *Sojourner* rover was the first wheeled rover to explore Mars. It worked for 84 days before it stopped working.

Picture of the *Sojourner* rover on Mars.

NASA sent 2 more rovers to Mars in 2003 named *Spirit* and *Opportunity*. They were sent to Mars to try and understand more about whether Mars once had flowing water. The spacecraft were sent to opposite sides of the planet.

They arrived on Mars in January 2004 and were designed to last for 90 days. In May 2009, the *Spirit* rover became stuck in soft soil. After that, *Spirit* continued to send NASA data though it was stuck in one spot. Later, NASA lost contact with *Spirit*, ending that mission after nearly 7 years of working on Mars.

Amazingly, *Opportunity* is still working today. That's over 10 years! Do you remember how long it was supposed to work? Only for 90 days! That's a well-built rover, don't you think?

NASA's newest rover landed on Mars in August of 2012. It is named *Curiosity*. This rover is about twice as long and 5 times as heavy as the *Spirit* and *Opportunity*.

An artist's drawing of the *Opportunity* rover on Mars.

*Curiosity* gathers and studies samples of rocks and soil. Its mission is to help scientists determine whether anything has ever lived on Mars. It's also studying the climate. Look at the picture of the different rovers. They are all different sizes. Which one do you think can perform more experiments? Why?

There are many more spacecraft exploring Mars. Some have landed on the surface, but cannot move around. Others are in orbit around the planet. One day, however, we hope to send astronauts to the Red Planet. Would you like to go?

*Opportunity* and *Spirit* rovers. Arrived in 2003. 400 lbs.

*Sojourner* rover. Arrived in 1996. 23 lbs.

*Curiosity* rover. Arrived in 2012. 2000 lbs.

## think about this

Humans often talk about traveling to Mars. We wonder about the possibility of finding life on other planets. Ray Bradbury, however said it best: "Today we have touched Mars. There is life on Mars, and it is us—extensions of our eyes in all directions, extensions of our mind, extensions of our heart and soul have touched Mars today. That's the message to look for there: We are on Mars. We are the Martians!" — Ray Bradbury, science fiction author, speaking at *The Search for Life* in our Solar System, a symposium at the Jet Propulsion Laboratory, Pasadena, California, 8 October 1976.

## What Do You Remember?

What is the name of the biggest volcano in our solar system? What is the atmosphere like on Mars? Describe the surface of Mars. What do you remember about the moons of Mars? How long does it take Mars to revolve and rotate? What is the weather like on Mars? Why do some astronomers think Mars would be a good place to visit and, perhaps, live? What makes Mars look red? What was your favorite part of this lesson?

LESSON 8

# SPACE ROCKS

## *wisdom from above*

When meteors streak across the night sky, these celestial fireworks are incredible to behold. Often, only the careful stargazer can see them. Some comets, however, leave trails for everyone to enjoy.

*Praise Him, sun and moon;*
*Praise Him, all stars of light!*

Psalm 148:3

Millions of space rocks orbit our sun. Some come from deep in our solar system. Others come from a region known as the asteroid belt, which is located between Mars and Jupiter.

# Comets

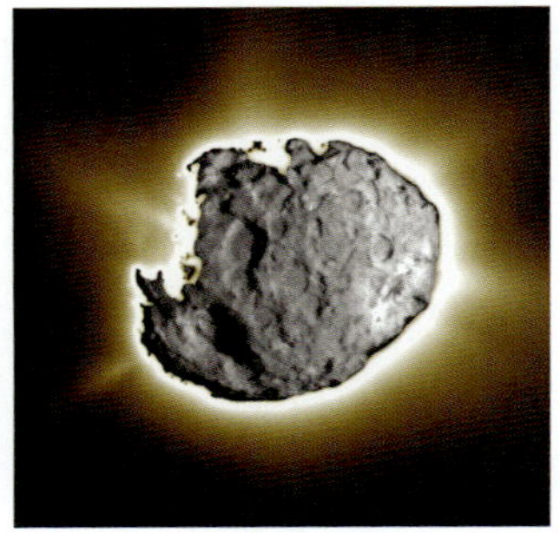

Have you ever tried to build a snowman when there was only a little bit of snow on the ground? If you have, you know that a lot of grass and dirt is gathered up with the snow. This makes a big, dirty snowball, which makes a big, dirty snowman! This is exactly what astronomers call comets—dirty snowballs! Of course, comets are much bigger than a little snowman and they don't have grass in them. In fact, many of them are so big that millions of snowmen could fit inside them. The illustration to the right is a NASA image of a comet called Wild (Vilt) 2. It was taken by a spacecraft called *Stardust*, the first NASA spacecraft designed specifically to study a comet.

Comet Hale-Bopp in 1997.

The name **dirty snowball** gives you a clue about what a comet is made of, doesn't it? Comets are mostly big balls of ice. The ice is a combination of frozen water and dry ice. This ice also includes a bunch of rock that forms the center of the comet, called the **nucleus** (new' klee us). If you could make a comet, you would collect lots of snow, ice, dirt, and rocks. Then, you would roll them all together into a giant dirty snowball. After that, you would toss it into space to orbit the sun! Of course, only God is mighty enough to make a dirty snowball large enough to travel through space.

The name comet comes from the Greek word *kometes*, which means head of hair. A comet looks like a star with a head of hair, or a big smudge in the sky. If you ever got paint on your finger and tried to rub it off on a piece of paper, it would make a smudge. That is what a comet looks like. It's a bright, white smudge of light. And though it looks small, it's really very large.

Comets have been seen for thousands of years. Written records confirm the observation of comets by ancient civilizations. Some comets were so bright that they could be seen during the day. Others had beautiful tails stretching halfway across the sky. Modern city lights outshine the stars making them difficult to see with the naked eye. This is called **light pollution**. Without light pollution, ancient cultures were able to see the night sky clearly and kept detailed written records of their observations.

Comet McNaught as captured by the European Southern Observatory in Chile. Both the comet and our sun were setting over the Pacific Ocean.

## *think about this*

In ancient times, before humans understood what space rocks were, many people believed that comets foretold the coming of evil. Believe it or not, some people still think that the night sky has power over their lives and their future! This is called **astrology**, and although it is similar to the word astronomy, it is very, very different! In astrology, people check their horoscopes to see what will happen to them.

God's plan for the stars is for His glory and not for man's! God's desire is for us to glorify Him and not the creation He made. If you ever meet anyone who follows astrology, God would have you share the truth about Jesus with him or her.

## The Coma

The reason a comet looks like a smudge in the sky is because of its **coma** (koh' muh). The coma is a big ball of steam and dust surrounding the ice. If you have ever seen dry ice, you have seen a gas that looks like steam coming off of the ice. That is similar to a comet's coma.

Comet Schwassmann-Wachmann 3 in 1995. Do you see how it looks like a big smudge in the sky?

When a comet first approaches the sun, it looks like a huge rock flying through space. As it gets close to the sun, it begins to heat up. The heat causes the ice to **sublimate**—that means the ice never melts into a liquid; it just goes straight from a solid to a gas which forms a cloud around the comet, making the coma. As the comet flies through space, it leaves a tail of this gas behind it. The coma can be very large and bright, but it always looks a little blurry since it is really gas. Now remember, the comet is orbiting the sun. Do you remember what solar wind is? It is a stream of particles shooting out from the sun. The solar wind pushes on the coma, and as a result, the coma always points away from the sun. Sometimes a comet has 2 tails; a yellow one and a blue one. The stuff in each tail reflects different light waves and produces different colors—the yellow tail is made out of dust particles, and the blue tail is the gas burning off the nucleus.

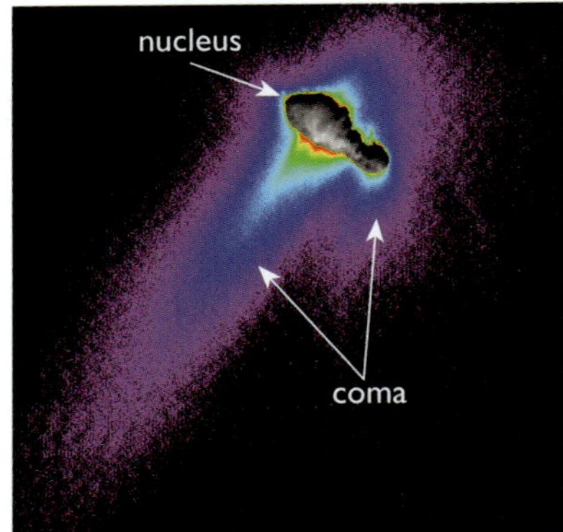


Comet Borrelly taken by the spacecraft Deep Space 1. False color has been used to emphasize the difference between the coma and nucleus.

If you have ever watched *Charlie Brown*, you might remember a character named Pigpen. A cloud of dust surrounds Pigpen, and when he walks, he leaves a trail of dust behind him. A comet is just like Pigpen! It always leaves a trail of dust particles behind it. These dust particles stay in the comet's orbit, so the comet has a very dusty, dirty orbit.

## A Comet's Orbit

Comets can have long, elliptical orbits that are far outside of the planets. Some orbit close to the planets, passing by Earth quite often, and some have orbits that take them far beyond our solar system for thousands of years. A comet always follows the same orbit, so it is easy to predict when a comet will pass by Earth again. Once a comet has been studied, scientists can predict when we will be able to see it again.

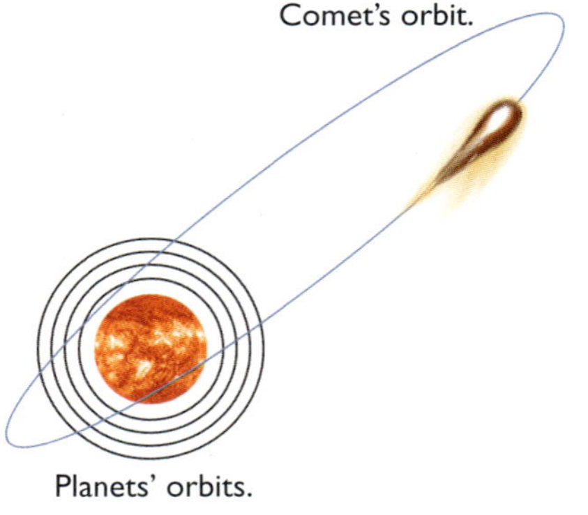


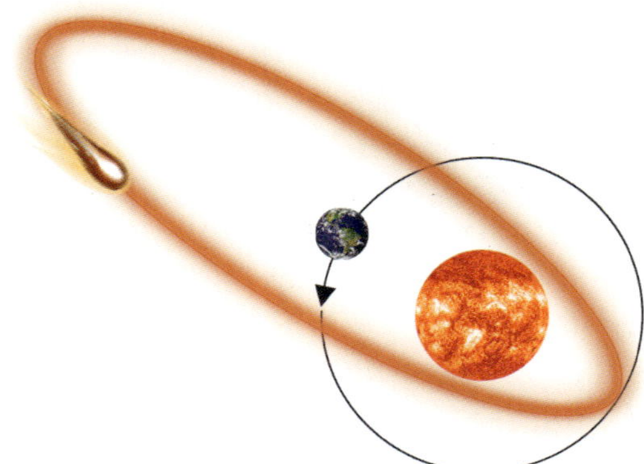

Remember that a comet is like Pigpen, leaving a trail of dirt behind it. When Earth passes through a comet's orbit, it runs into the dust and dirt that the comet leaves behind. An amazing and spectacular thing happens when these dust particles hit Earth's atmosphere—they catch on fire and create a little fireworks display for us! As they burn up, they look like beautiful shooting stars. It's a sight to see! These particles are very small, but because they burn so brightly when they hit our atmosphere, they appear the same size as a star far out in space. Many people believe that a shooting star is an actual star up in space, but it's not. It's just a little piece of comet. When Earth moves into that dusty, dirty comet path, we see numerous shooting stars. When we have shooting star shows like that, we call them **meteor showers**.

# Activity 8.1

## Watch a Meteor Shower

| Shower Name | Approximate Date to See | Name of Comet that left its Dust (if known) | Number of Shooting Stars per Hour |
|---|---|---|---|
| Quadrantids | January 3 | | 40 |
| Pi Puppids | April 5 | | 40 |
| Lyrids | April 22 | Comet Thatcher | 15 |
| Eta Aquarids | May 5 | Comet Halley | 20 |
| Delta Aquarids | July 30 | | 20 |
| Perseids | August 12 | Comet Swift-Tuttle | 50 |
| Orionids | October 22 | Comet Halley | 25 |
| Taurids | November 4 | Comet Encke | 15 |
| Leonids | November 17 | Comet Temple-Tuttle | 15 |
| Geminids | December 14 | Asteroid 3200 Phaethon | 50 |
| Ursids | December 23 | Comet Tuttle | 20 |

Find the meteor shower that is closest to today's date. Then go to **www.apologia.com/bookextras** and find the exact date or dates to view the meteor shower this year. Mark your calendar, and then go out and watch the show! Here are some tips on how to best view a meteor shower:

1. Do not use binoculars or a telescope. Your eyes are the best instruments to use when watching meteor showers.
2. Try to get away from city lights that drown out the meteors. If you live in the city, see if you can go to a park or a country field.
3. Try to stay out later than you normally would. Typically, the best times to watch meteor showers are between midnight and dawn. If you can't stay up that late, go out when it is very dark.
4. Bring a blanket or lawn chair that will recline. Lie on the ground or on the lawn chair and look up. Let your eyes adjust to the dark sky, and just keep watching for streaks of light in the sky. Those are the meteors.
5. Count how many you see. See if the number comes close to the number on the chart.

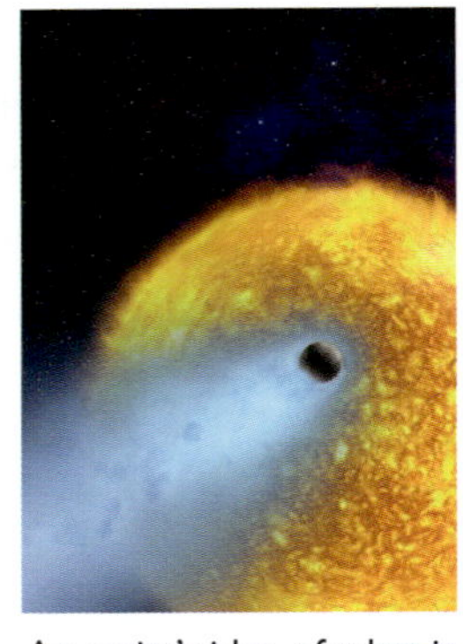
An artist's idea of what it looks like when a comet passes close to the sun.

Because we know where the comets's orbits are, we know when Earth will pass through them. As a result, we can predict when we will have a meteor shower. Notice that on or around August 12th each year, the **Perseid** (per' see id) meteor shower—named for a constellation of stars that appears in the sky at that time—gives us 50 shooting stars every hour. That is one you don't want to miss, so mark your calendar now!

If a comet's orbit around the sun takes less than 200 years, it is called a **short-period comet**. While 200 years might not seem like a very short period to you and me, for a comet, it is quite short. After all, compare that to **long-period comets** that can take thousands of years to make one orbit around the sun!

Some short-period comets get heated up by the sun so often that the ice around the comet gets burned off. It orbits the hot sun too often to stay a big ball of ice for very long. Suppose you kept bringing a dirty snowball inside by the fireplace every few minutes. What would happen? It would eventually lose the snow and be a pile of dirt. Since there is rock in all of that ice, when a comet's ice burns off, it looks like a big rock floating in space, and is called an **asteroid** (as' tuh royd).

**Take a moment to review with someone else all that you have learned so far about comets. What are they made of? What are their parts called?**

# Famous Comets

One famous short-period comet is known as comet Halley, or Halley's Comet. This comet appears every 76 years. Chinese stargazers saw it 240 years before Christ was born! The Chinese were very interested in the universe and kept great records of all they saw in the sky. In the 1600s, Edmund Halley studied ancient records and determined that the comet recorded by ancient Chinese astronomers returns every 76 years. Isn't it amazing that we can study ancient documents, written thousands of years ago, and learn more about astronomy? Sadly, Edmund Halley never got to see the comet's return and confirm that he was right. However, the comet did appear, just when Mr. Halley predicted it would. Because of this, we named the comet after him. Do you think you will ever have a comet named after you? What would we call the comet you discovered?

Close view of Halley's Comet. You can see the coma forming around the central nucleus.

Comet Hale-Bopp.

The biggest and brightest comet seen from Earth in the last century is comet Hale-Bopp. Its name is a little strange because 2 men discovered it, so it is named after both of them. This is a long-period comet, which means it will not return to Earth for quite some time. It is so gigantic that its coma is as big as the sun. It didn't come close enough to the sun to cross our orbital path, so we did not get the pleasure of seeing a meteor shower display from Hale-Bopp's dust trail. Of course, it's probably a good thing that it did not pass too closely to Earth. A close encounter with such a big comet would be scary. Imagine what a comet larger than the sun would look like if it got too close to Earth!

The Shoemaker-Levy 9 comet made history when it came too close to Jupiter. While orbiting Jupiter in July of 1992, this comet broke into several comets. Two years later, in July of 1994, an amazing thing happened. As the pieces of Shoemaker-Levy 9 were once again approaching Jupiter, they were pulled right into the planet by its gravity. They crashed into Jupiter for 6 days, causing giant explosions on the gassy surface of Jupiter. It was an amazing sight to see.

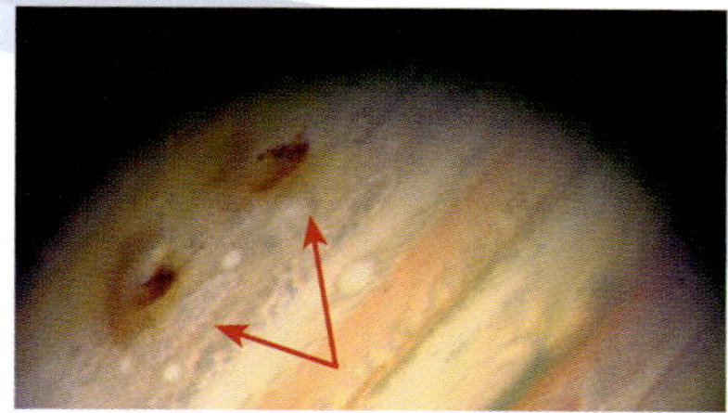

This Hubble Space Telescope image of Jupiter shows 2 spots where pieces of Shoemaker-Levy 9 hit.

If a comet comes by Earth, be sure to take a good look at it because it probably will not return in your lifetime. Even though you will not see the comet again, you may be able to see its trail every year when Earth passes through its path of dirty particles. Comets are beautiful lights traveling through the sky, and the trail of dust they leave behind gives us an amazing shower of lights. All of this reminds us of the glory of God and how He formed the world to show us how great He is. These beautiful gifts of celestial lights reflect His bright, shining glory.

## How Comets Get Their Official Names

Comets are usually named for their discoverer—either a person or even a spacecraft. For example, comet Shoemaker-Levy 9 was the ninth short-periodic comet discovered by Eugene and Carolyn Shoemaker and David Levy.

A photograph of meteors during the Leonid meteor shower.

A meteorite which scientists believe was once a part of Mars. It probably launched from Mars as a result of a space rock impact before landing on Earth.

## Meteoroids, Meteors, Meteorites

Remember all the dirt that comets leave behind? In addition to the dirt and rocks left by comets, dirt and rocks from planets float about in space. We call them **meteoroids** (mee' tee uh royds'). Each time a meteoroid hits Earth's atmosphere and burns up, it is called a **meteor** (mee' tee or). As you already learned, meteors burn up inside our atmosphere and look like little streaks of light, which we often call shooting stars. If a very large meteor hits our atmosphere, it will burn very brightly. These bright meteors are often called fireballs.

Most meteors are very small. But occasionally, an object is so large that when it falls through Earth's atmosphere, it does not completely burn up. When these objects hit Earth's surface, they are called **meteorites** (mee' tee uh rites'). Usually, meteorites fall into water, since Earth is mostly covered by oceans. Once in a while, however, a meteorite will fall onto land. Some people are meteorite hunters who spend all their free time searching Earth for meteorites. The meteorites that have been found on Earth have been many different sizes—some very small and others very large. The largest one found weighed 60 tons. That's as heavy as a whale, or 30 cars! Most meteorites are either iron, stony or stony-iron.

There is a place in Antarctica where the ground is covered with meteorites. More than 10,000 of them have been found lying on the snow. Antarctica is cold, and the ground there is always frozen with multiple layers of ice. Since the rocks were sitting above the solid ice, astronomers know these rocks fell from the sky. Several of them are made of the same material found on Mars, so scientists think that they are Martian meteorites.

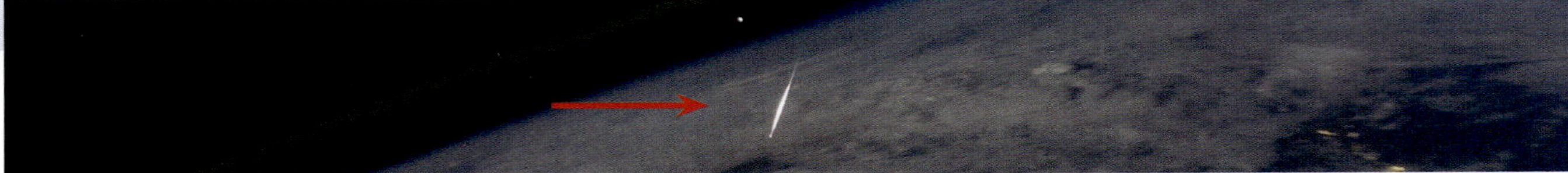

This photo was taken from the International Space Station by a NASA Astronaut showing a meteor over Earth.

**Stop for a moment and tell someone what you know about the difference between meteoroids, meteors, and meteorites.**

Meteor Crater (also called Barringer crater) is one of the youngest and best-preserved impact craters on Earth. It is located in the Arizona desert and is about 1 kilometer wide (0.6 mile). There are about 170 impact craters identified on Earth.

## Asteroids

An asteroid is a rock orbiting the sun in our solar system. Wait a minute. Isn't a rock orbiting the sun called a *meteoroid*? Well, yes, it is. The difference between an asteroid and a meteoroid is size. **Asteroids** are large rocks (typically larger than a football field), while meteoroids are small rocks. Although some asteroids are large, even if you combined all the asteroids together, the object would be smaller than Earth's Moon!

Asteroids are made of iron, rock, and carbon. When they enter our atmosphere, they burn up, just as meteoroids and comets do. When an asteroid hits Earth, it is called a meteorite. However, it looks so much like Earth's rocks that unless it falls in a place where there are not very many rocks it is hard to tell it apart from other rocks. The ones that are found most often are the iron ones because they are shiny and you can pick them up with a magnet. There could be a meteorite in your backyard, but you probably can't tell the difference between it and a rock from Earth, unless you have a magnet.

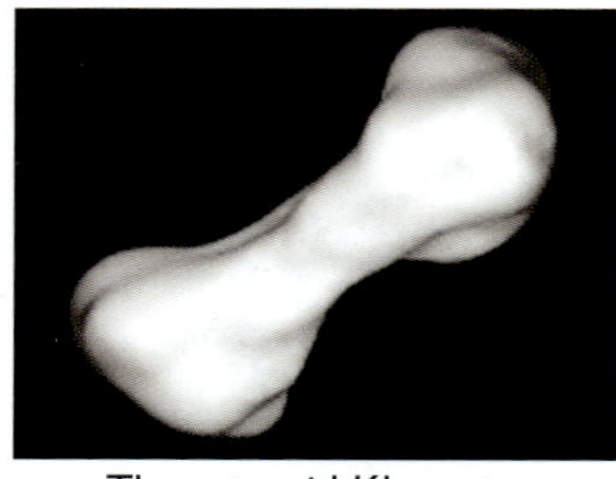

The asteroid Kleopatra.

There are millions of asteroids out in space. Some are the size of a football field; some are bigger than the state you live in. They come in all shapes as well. Many of them have been named, like Kleopatra, an asteroid that looks like a big dog bone! Asteroids are sometimes described as mountains in space or **planetoids**, meaning little planets. Most of them can be found between Mars and Jupiter, in what is known as the **asteroid belt**. When a spacecraft goes to explore one of the planets past Mars, it must first pass through the asteroid belt. Fortunately, the asteroids are spaced far enough apart that the spacecraft can steer through them.

**Can you tell the differences between comets, meteoroids, meteorites, and asteroids?**

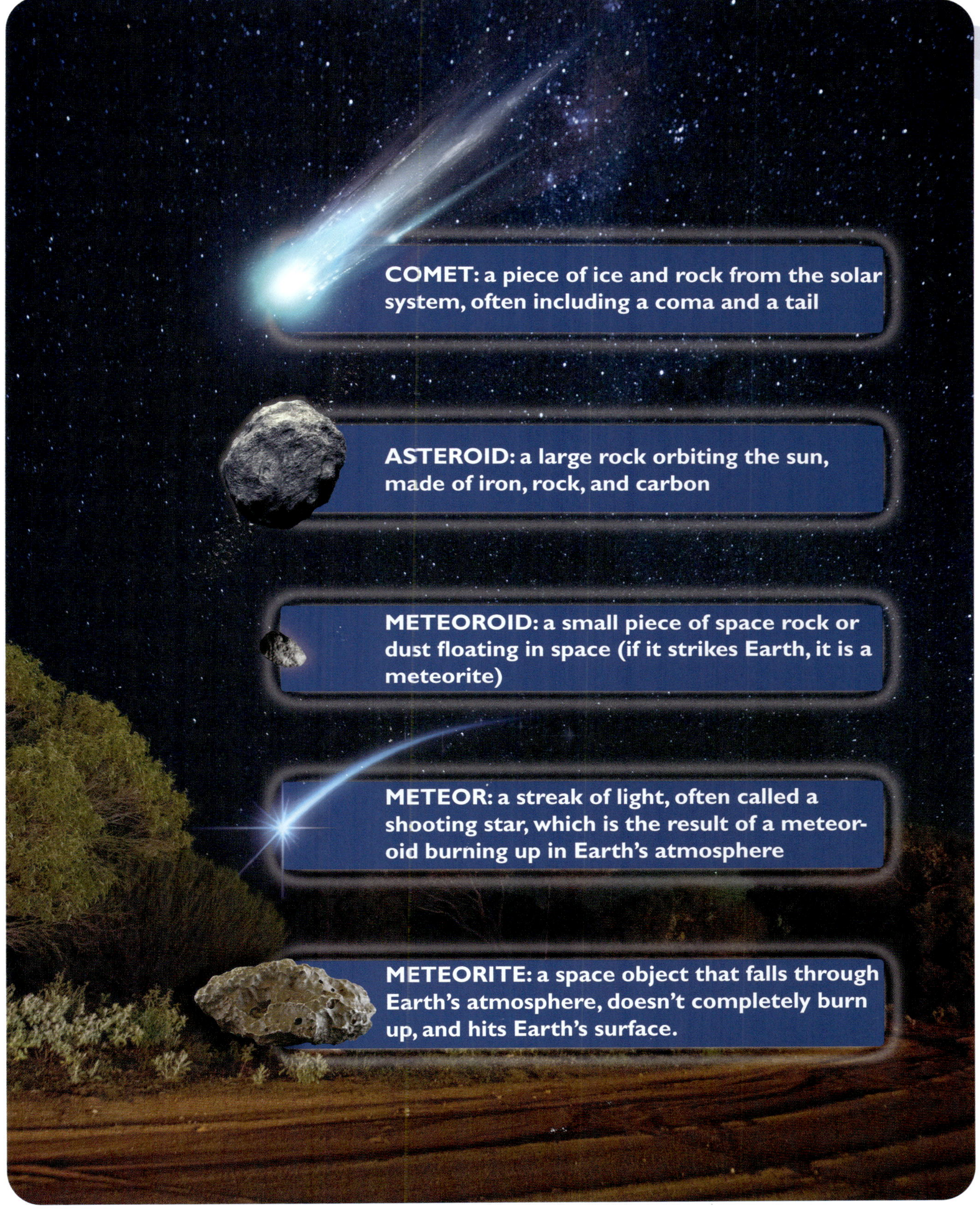
COMET: a piece of ice and rock from the solar system, often including a coma and a tail
ASTEROID: a large rock orbiting the sun, made of iron, rock, and carbon
METEOROID: a small piece of space rock or dust floating in space (if it strikes Earth, it is a meteorite)
METEOR: a streak of light, often called a shooting star, which is the result of a meteoroid burning up in Earth's atmosphere
METEORITE: a space object that falls through Earth's atmosphere, doesn't completely burn up, and hits Earth's surface.

# Asteroid Belt

Let's learn a little more about the asteroid belt. The asteroid belt is a ring of asteroids that orbit the sun between Mars and Jupiter. We actually use the asteroid belt to separate the planets into 2 groups. Planets that are inside the asteroid belt are closer to the sun and are called the **inner planets**. Planets outside the asteroid belt are called the **outer planets**. Mercury, Venus, Earth, and Mars are the inner planets, while Jupiter, Saturn, Uranus, and Neptune are the outer planets.

In the drawing above, you will see that the asteroid belt looks a lot like the orbit of a planet. Well, there may be a reason for this. There is evidence that there was once a planet that orbited the sun between Mars and Jupiter. What happened to that planet? It may have exploded, and the asteroids in the asteroid belt may be the remains of that planet.

The idea that there was once a planet between Mars and Jupiter is called the **Exploded Planet Hypothesis** (hi pahth' uh sis). A hypothesis is an idea that may or may not be true. When something is a mystery, you look at clues and form an idea of what you believe. This is a hypothesis! For example, suppose you left a cookie on the table. If you came back to the table later and found your cookie was gone, you would have to form a hypothesis about its disappearance. Did someone eat it? Did Mom throw it away? Did Dad put it away? Perhaps someone opened the door, letting the wind blow in so strongly that it blew your cookie off the table, and then a giant ant came and took it back to its ant hole. As you can see, some hypotheses are more likely than others.

There are many reasons why the Exploded Planet Hypothesis is a good hypothesis. If a terrestrial planet exploded, the pieces flying into space would be rocks and ice. That is exactly what asteroids and comets are! Our planet is made of mostly water, so if it exploded, it would send millions of tons of water that would become ice out in space, perhaps forming huge comets. Scientists think that some of the comets and asteroids in our solar system were created when this former planet exploded. Did the water from this former planet form ice around the rocks when they were ejected into freezing cold space? We don't know for sure, but it supports the Exploded Planet Hypothesis.

More evidence supporting the Exploded Planet Hypothesis are the crater scars on many planets and moons. Remember that when a giant meteorite hits, it leaves a big dent in the surface. Some scientists believe that meteorites hit the planets and moons randomly over billions of years. But when one side doesn't have very many crater scars, as is true of the planets in our solar system, it really bewilders them. Let me try to explain why.

Suppose you hung a big ball of clay from the ceiling and started it slowly spinning. Then, suppose you walked around and around the ball, throwing marbles at it for a year. If you did that, you would find dents all over the ball. If you had 10 balls spinning and you walked about throwing marbles at them for many years, the dents would be spaced out randomly over all the balls. The longer the balls hung while you threw marbles at them, the more likely it would be that craters would be evenly spaced all over the balls. On the other hand, suppose you threw marbles for only a few minutes. If you did that, there would not be enough marble dents to be evenly spaced on each ball. Most likely, they would be more concentrated on one side than the other.

Now suppose you had all 10 balls slowly spinning around, and one ball suddenly exploded into pieces that went violently and forcefully flying about the room. Most of the other balls would get hit, but they would mostly be hit on the side facing the exploding ball. Furthermore, the balls closest to the exploded ball would probably have the deepest dents.

This is exactly what we see with our planets and moons! They have a great many more dents on one side than any other. This is true of Earth and its Moon, Mars, Mercury and many of the moons orbiting Jupiter, Saturn, Uranus and Neptune. In fact, the planet closest to the asteroid belt, Mars, has extremely deep craters. There is a crater on Mars, called Hellas Basin, that is one-third the size of the United States of America! If the asteroid belt was a planet that exploded, you would expect Mars to have had the most violent meteorite hits, resulting in the largest craters—and it does!

Do you recall that pieces of Mars have been found on Earth? We call them Martian meteorites. How did those get to Earth? If the planet next to Mars exploded, gigantic chunks of that planet would have hit Mars at such a high speed that it could have caused pieces of Mars to launch into space.

An image of Mars, showing the huge Hellas Basin. The colors are false; they are used to illustrate differences in the surface of the planet.

An artist's idea of a large asteroid colliding with a planet.

Now, of course, if a planet exploded, pieces of it would certainly hit Earth as well. They would probably hit Earth at the same time that the Martian meteorites were hitting Earth because they all formed as a result of the same explosion. Do you remember the thousands of meteorites I told you about in Antarctica? Of the thousands of meteorites there, only a few are from Mars. It's possible that most of those meteorites hit Earth as a result of an exploding planet. We might never know how these other meteorites got to Earth, but it is the job of scientists to search for answers.

Some scientists say that the Exploded Planet Hypothesis couldn't be true because there are not enough asteroids in the asteroid belt to make up a whole planet. However, if we combined all the comets and other planetary debris in our solar system, the exploded planet would be much bigger than Earth!

If the Exploded Planet Hypothesis is correct, why did the planet explode? Some scientists think it was because a huge volcano erupted. Yellowstone National Park is built upon a gigantic, super volcano. This volcano is so big that if it erupted, at least 3 states would have to be evacuated! See if you can find Yellowstone on a map of the United States (look in the northwest corner of Wyoming).

I doubt that the super volcano under Yellowstone will erupt in your lifetime, but if it did, it would mean big trouble. It would probably not cause Earth to explode, but it would cause a lot of damage. If a planet once existed where the asteroid belt is now, and if that planet had a super volcano that was even larger than the one under Yellowstone National Park, it is possible that such a large eruption could have caused the planet to explode.

Of course, if this planet once existed, there is another possible cause for its explosion. Some believe the planet between Mars and Jupiter exploded because a gigantic object, maybe a huge comet or asteroid, hit the planet. If a big asteroid or comet can cause huge craters such as the Hellas Basin on Mars, an even bigger asteroid or comet could cause a crater so deep that it would tear an entire planet apart!

The asteroid belt is a very fascinating part of the solar system. It is a wonderful mystery that is fun to try and solve.

**Use your own words to describe the asteroid belt.**

## Who Names Asteroids

You've learned that many planets were named for Roman gods. You've also learned that many comets are named after people who discovered them. With asteroids it's different, and sometimes even a bit fun. There is an asteroid named Spock (from "Star Trek") and another named Zappa (a rock musician). They can be named for places or just given a number. Some asteroids are named in tribute, such as the seven asteroids named for the crew of the Space Shuttle Columbia killed in 2003.

## Spacecraft

Several NASA, European, and Japanese unmanned spacecraft missions have flown by and observed asteroids. Some have landed on asteroids. Some have even returned samples of asteroid dust back to Earth so that scientists can learn more about these fascinating space rocks.

# Activity 8.2

## Understanding Distances in Our Solar System

If our sun was smaller than a dime, the following distances would be a good representation of planetary distances.

### You will need:

- Paper
- Markers
- Tape
- A large open space

### You will do:

1. Write one name on each sheet of paper for Sun, Mercury, Venus, Earth, Mars, Astroid Belt, Jupiter, Saturn, Uranus, Neptune, Kuiper Belt. You can get creative and draw the planets as well.
2. Find a large open space that can accommodate at least 100 steps. Each step should be a little over 3 feet.
3. Take your papers and tape them to the ground as follows:

    Sun: at the edge of the area
    Mercury = 1 step from sun
    Venus = 2 steps from sun
    Earth = 2.5 steps from sun
    Mars = 4 steps from sun
    Asteroid belt = 7 steps from sun
    Jupiter = 13 steps from sun
    Saturn = 24 steps from sun
    Uranus = 49 steps from sun
    Neptune = 76 steps from sun
    Kuiper belt = 100 steps from sun

### Discussion

In this activity, you know that the distances are fairly accurate if the sun is smaller than a dime. Can you imagine the distances you would need if the sun were the size of a basketball? How about the size of your home? Think about the distances with the sun its actual size. That's a lot of space!

Because distances are so great in our solar system, scientists use what is called an astronomical unit, or AU, to describe them. An AU calculates a planet's distance from the sun based on the Earth's average distance from the sun (over 90 million miles). The Earth is the only planet in our solar system that can be AU 1.

What do you think the AU for the other planets would be? The planets closer to the sun would have an AU smaller than 1; the planets farther out from Earth would have an AU greater than 1.

## think about this

Have you ever watched a movie that took place in outer space? Sometimes the movies make the asteroid belt look like it is filled with rocks and debris. This is not true. Asteroids are very far apart from one another. On average, asteroids are about 1.2 million miles apart from each other.

## What Do You Remember?

What is another name for a comet? What does a comet leave behind as it orbits the sun? What happens when a comet's dust particles enter our atmosphere? What do people call meteors? What is a meteor called when it hits Earth? Where have many meteorites been found? From which planet did some of the meteorites come? Where is the asteroid belt located? What is the Exploded Planet Hypothesis? Can you give some reasons that this might be a correct hypothesis? What was your favorite part of this lesson?

## LESSON 9
# JUPITER

## *wisdom from above*

Just like a mother protects her children, Jupiter protects Earth. But where God gave mothers love to keep children safe, He gave Jupiter a large mass to keep Earth safe!

*Your lovingkindness, O Lord, extends to the heavens, Your faithfulness reaches to the skies.*

Psalm 36:5

Jupiter and its Great Red Spot.

# Protective Mother

Jupiter is the first of the outer planets, and it is the giant of our solar system. It's the biggest planet that orbits the sun. It is so big that all of the other planets could fit inside of it. That's gigantic! If Earth were the size of a marble, Jupiter would be the size of a basketball.

Jupiter is like a protective mother to Earth. You see, Jupiter is so massive that it has a very strong gravitational pull. Do you remember we talked about the rocks flying around in space? Meteoroids hit Earth and burn up, but comets and asteroids are much, much larger than meteoroids. We do not want any comets or asteroids flying into Earth! Thankfully, Jupiter's gravity pulls so strongly that most comets and asteroids eventually hit Jupiter instead of Earth. Do you remember the Shoemaker-Levy 9 comet you learned about in lesson 8? Its pieces hit Jupiter when it broke apart. They did not hit Earth. Just think what might have happened had Jupiter not been there! We can thank God for making Jupiter so big that the comets do not fly into Earth. Life on Earth would be very difficult if it were not for Jupiter's large gravitational pull!

# Going to Jupiter

Jupiter is the planet just past Mars, but if you wanted to go to Jupiter, you would have to pass through the asteroid belt in order to get there. That might be dangerous. When astronomers send spacecraft to Jupiter, it is a tricky thing to get the spacecraft past all those flying boulders.

Arriving at Jupiter does not solve all the difficulties of visiting this planet. If a spacecraft tried to land on Jupiter, it would sink right down to the very core of the planet. There isn't any ground to stand on! That's because Jupiter is not a terrestrial planet. It's what we call a **gas giant**. It has a small, rocky core, but all around this small core are swirling gases. This means that Jupiter is primarily a big ball of atmosphere. The atmosphere is definitely not pleasant—it's a very dangerous and hostile atmosphere. We definitely could not breathe if we went to Jupiter. Also, radiation levels encountered by spacecraft visiting Jupiter are more than 1,000 times the

Jupiter is 484 million miles away from the sun.

lethal (deadly) level for a human. Even heavily-shielded spacecraft like *Galileo* were damaged by the radiation. If we could space travel, we would not want to visit Jupiter for very long unless we had the right equipment to keep us alive.

## Little Sun

Jupiter is too far away from the sun to get much heat from it. As a result, it is a cold planet. However, it's not as cold as you might think because Jupiter has its own heat source. Remember, Jupiter has a small, rocky core. It turns out that this core is very hot, and it cools off when heat escapes into the surrounding gases. Instead of just absorbing the sun's heat, Jupiter also produces its own heat! Although the heat put out by Jupiter is tiny compared to the heat of the sun, Jupiter is like a little sun.

Jupiter has something else that makes it similar to the sun. The sun is made mostly of 2 gases, **hydrogen** (hi' druh jen) and **helium** (he' lee uhm). You might recognize helium; it is the gas that makes balloons float in the air. If you let go of a balloon filled with air, it falls to the ground. If you release a balloon full of helium, it floats into the sky. Like the sun, Jupiter is also made mostly of hydrogen and helium. In addition to putting out heat like the sun, and being made of the same gases as the sun, Jupiter even has natural satellites like the sun. The sun's natural satellites are the planets. Jupiter's natural satellites are its moons.

## Stormy Skies

Most pictures of Jupiter show a large spot on the planet's surface. That's the **Great Red Spot**, a giant storm (like a hurricane) that travels around the planet. And great it certainly is; the Great Red Spot is twice the size of Earth! This storm is many times stronger than the worst hurricane on Earth, and it has been raging on Jupiter for more than 300 years! How do we know this? Astronomers have seen it in their telescopes since at least 1665. That is one huge, long-lasting storm.

In pictures, we see beautiful stripes crossing Jupiter. However, those stripes aren't all that beautiful when you get close. Those stripes are clouds of violent storms. Jupiter's lightning is much more powerful than the lightning on Earth. With all of those storms and lightning going on, Jupiter seems like a very frightening place. It's hard to imagine how we could ever send people to Jupiter.

# Activity 9.1

## Make a Hurricane Tube

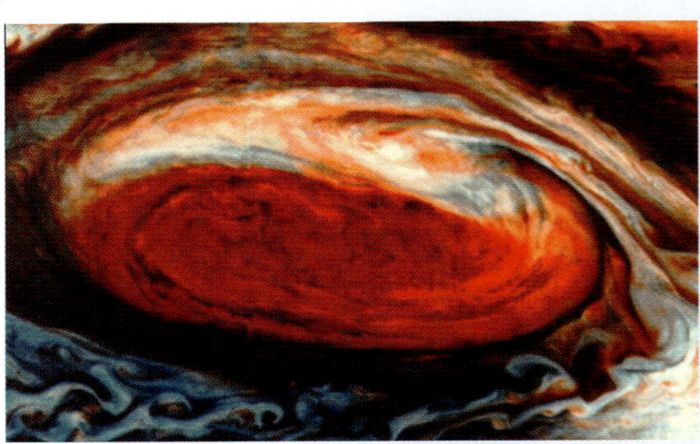

A close-up view of Jupiter's Great Red Spot.

### You will need:

- Adult supervision
- 2 plastic soda pop or water bottles (Larger ones give you a bigger hurricane.)
- 1-inch washer
- Water resistant tape (Electrical or duct tape works best.)

### You will do:

1. Fill one bottle two-thirds full of water.
2. Place the washer on top of the mouth opening (see the picture on the top right).
3. Place the other bottle upside-down on top of the washer.
4. Tape the 2 bottles together very thoroughly (see the picture on the bottom right). If your experiment leaks, you will need to add more tape and wrap it tighter.
5. Turn the 2-bottle setup over so that the bottle with water in it is now on top.
6. Move the bottles in a circular motion, swirling the water around.
7. Watch your system form a hurricane-shaped funnel down into the bottle below. You can do this over and over again. Try putting glitter or tiny toys in the hurricane tube and watch them spin down into the tube below.

**Tell someone everything you have you learned about Jupiter so far so that you won't forget it.**

# Jupiter's Rings

We have always known that Saturn has rings, but astronomers have recently discovered that Jupiter has rings around it, too! These rings are not visible through telescopes because they are very thin, and the light that Jupiter reflects from the sun makes them difficult to see. In fact, they were not seen until the *Galileo* orbiter got close enough to take a picture of them. We usually don't illustrate Jupiter with its rings since you cannot see them with a telescope.

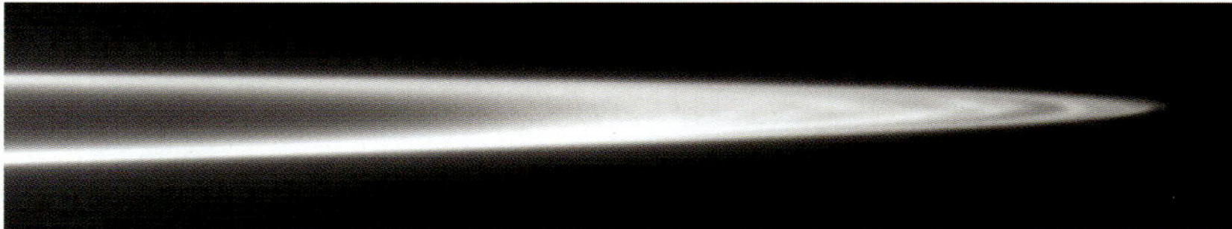

Jupiter's rings as photographed by *Galileo*. Jupiter cannot be included in the picture because the light from Jupiter would drown out the light being reflected by the rings.

## Rotation and Revolution

Jupiter rotates very quickly. It takes only 10 hours for Jupiter to rotate one time. That means that a day on Jupiter is less than half of an Earth day. Nighttime on Jupiter is only 5 hours, and daytime is 5 hours as well. Those are short days and nights! Do you remember from lesson 5 that the speed of a planet's rotation affects its weather? Earth's rotation is slow enough that the weather is not very violent. Jupiter's fast rotation is part of the reason there are violent winds and storms there.

Jupiter is further away from the sun than Earth, so a revolution around the sun takes much longer. It takes Jupiter 12 Earth years to orbit the sun. We call that one Jovian (joh' vee un) year. The word *Jovian* refers to Jove, the Roman god of rain, thunder, and lightning. Another name for this false god is Jupiter, so *Jovian* means Jupiter. Since one Jovian year is 12 Earth years, when you turn 12 on Earth, you would be only one Jovian year old. You would be a big one-year-old wouldn't you? How old would you be on Jupiter if you were 48 Earth years old?

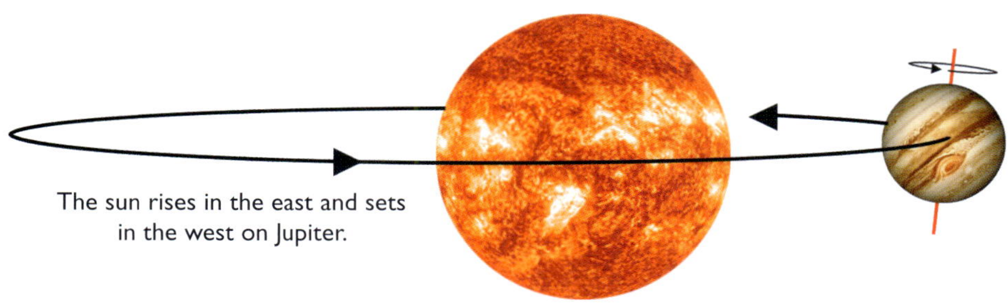
The sun rises in the east and sets in the west on Jupiter.

## Many Moons

Do you remember that a moon is a natural satellite? Well if Jupiter is like a mom, she has a lot of children! Jupiter has more than 50 moons with an additional 17 moons awaiting confirmation of their discovery—that is a total of 67 moons. It seems that most of Jupiter's satellites are adopted. While many objects pulled by Jupiter's strong gravity crash into the planet, it appears that some became satellites instead.

The 4 biggest satellites, or moons, orbiting Jupiter are named **Io** (ee' oh), **Europa** (yuh roh' puh), **Ganymede** (gan' uh meed), and **Callisto** (kuh lis' toh). Galileo discovered these moons many years ago. Do you remember who Galileo was? He was a famous astronomer who taught us a great deal about space. He was the first to use a telescope to study the heavens in depth. In honor of Galileo's discovery, these 4 moons are called *Galilean* moons.

Europa is the smallest of the Galilean moons. It does not have craters like the other moons. Astronomers are amazed at how smooth Europa is. In fact, it is the smoothest object in the solar system. Most astronomers think that Europa is covered with a large, frozen ocean. Do you see all of the lines in the photograph of Europa? Those are cracks in the ice. Evidence from the Hubble Space Telescope tells us that there is oxygen on Europa. However, the amount of oxygen on Europa is so small that it could not support life.

A size comparison of Jupiter and the Galilean moons.

Europa - an icy moon.

Io - a colorful moon.

Io is a strange name isn't it? It is the second smallest of the 4 Galilean moons. Io has hot, active volcanoes, but they aren't filled with lava! They are filled with a sulfur chemical that smells like rotten eggs. Every once in a while, NASA astronomers see the sulfur chemical spewing into the atmosphere of Io, and that's what gives Io its interesting color. In fact, Io is the most colorful moon in the solar system.

Jupiter's second largest moon, Callisto, is about the same size as the planet Mercury. Astronomers recently learned that Callisto is a strange moon, indeed. It doesn't have any core at all, the way most moons and planets do. It is a huge ball of ice with rocks and boulders embedded in it. It's very much like the nucleus of a comet. Interestingly enough, Callisto has one of the largest multi-ring impact craters in the solar system. It is called Valhalla, and its rings extend about 2400 miles across. That must have been one big asteroid or comet!

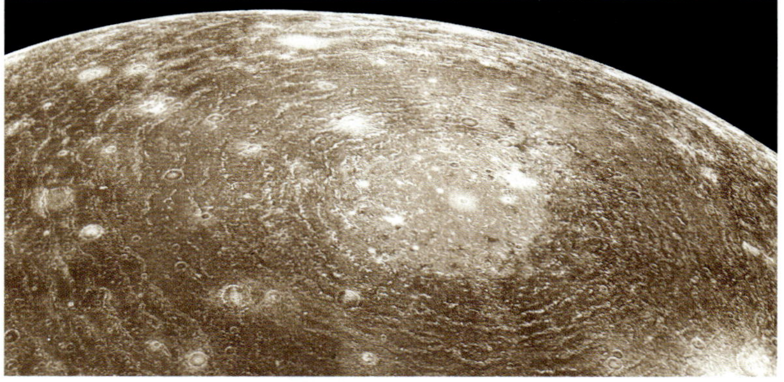
Callisto's multi-ring impact crater.

Ganymede is Jupiter's biggest moon. In fact, it is the biggest moon in our whole solar system. It's bigger than the planet Mercury. It would be considered a planet if it were circling around the sun instead of Jupiter. It is similar to Callisto in the sense that it is made of ice and rock, and covered with craters. Unlike Callisto, however, it is not just a simple mixture of ice and rock resembling a comet's nucleus. Instead, this moon has a definite structure. There is a hot core made of iron and sulfur, a rocky covering around the core, and an icy covering above that. Like Europa, there is some oxygen on Ganymede, but it is definitely not enough for you to breathe if you wanted to visit there. Ganymede is the only moon known to have its own internally generated magnetic field.

It seems that many of Jupiter's moons are really gigantic balls of ice. Some believe these frozen ice balls were comets pulled into orbit by Jupiter's gravity.

Ganymede - a giant moon.

**Amalthea** (um al thee' uh), the reddest object in the solar system, is a very odd Jovian moon. It is a jumble of gigantic rocks pulled together by gravity. Suppose you grabbed a bunch of rocks from your yard and glued them together to make a rock structure. Your rocks wouldn't fit perfectly together because they were not made to go together. That's like Amalthea. It has big spaces between the boulders, just as your rock structure would have. Your rock structure probably wouldn't be round. Amalthea isn't round either. It is also small compared to the Galilean moons. Notice how small it is compared to Io, the second smallest Galilean moon. Amalthea might be pieces of the exploded planet that formed the asteroid belt.

A size comparison of Io and Amalthea (upper right).

**What do you know about Jupiter's moons? Explain what you have learned.**

Astronomers use this symbol for Jupiter.

## Finding Jupiter in the Night Sky

Jupiter is the second brightest planet in our solar system. Jupiter is so large that you can see some of its details, even its moons, with just a pair of binoculars or a small telescope! The way to see the moons with binoculars is to steady the binoculars against something solid like a car or a mailbox. This will keep them focused on Jupiter. After your eyes have had a minute to adjust to the light, you will see one or more small specks of light around Jupiter. Those are Jupiter's moons!

If you need help finding Jupiter in the sky, visit **www.apologia.com/bookextras** where you will find links to helpful websites.

## Who Named Jupiter

Jupiter was named after the Roman's king of gods, Jupiter, who was also the god of the sky and of thunder.

## Spacecraft Galileo

Almost everything we know about Jupiter comes from the unmanned spacecraft called *Galileo*. In 1989, *Galileo* began an interesting journey to Jupiter, taking 6 years to get to there. It would not normally take that long to get to Jupiter, but this spaceship took the long way around.

An artist's rendition of *Galileo* reaching Jupiter and beginning its study of Io.

When *Galileo* left Earth, it traveled away from Jupiter, toward Venus. Why? NASA engineers often use the gravity of planets to propel spacecraft into outer space. They call it **gravity assist**. When a spacecraft gets close to a planet, the planet's gravity grabs the spaceship. With all the speed coming from the spaceship, the pull of the gravity, and the circular motion of the orbit around the planet, the spaceship gets a big push into outer space without using precious fuel. *Galileo* used Venus's gravity to swing around the planet and fly toward Jupiter. When it flew past

The Space Shuttle Atlantis carrying the *Galileo* orbiter and atmospheric probe as it lifts off from Kennedy Space Center.

Earth, *Galileo* took some beautiful pictures of our planet and got a big swing from Earth's gravity, too. Mars was on the other side of the sun at the time, so *Galileo* didn't get any pictures. After it passed Mars's orbit, *Galileo* approached the asteroid belt. Thankfully, it didn't hit any asteroids and no asteroids hit it—and it took some pictures of passing asteroids. Since it was traveling on a curved path, *Galileo* came back to Earth and got one more big push into outer space and was finally on its way to Jupiter. *Galileo* again passed through the asteroid belt, without any collisions, and arrived safely at Jupiter.

*Galileo* had one big problem when it got out into space. Its big antenna would not open! NASA needed a strong antenna to send information across the solar system from Jupiter. A little antenna cannot send many pictures or much information. The astronomers on Earth tried everything to get the antenna to come up. They tried pointing *Galileo* into the sun; they tried turning it away from the sun; they even tried turning *Galileo* on and off. They tried everything, but nothing worked. With only the small antenna, they did not get as much information as they had hoped. Still, *Galileo* took many beautiful photos while on its journey. In fact, most of the pictures in this lesson come from *Galileo.*

When *Galileo* got to Jupiter, it sent a probe into the middle of Jupiter to gather information. The probe was released from the main spacecraft in July 1995, 5 months before reaching Jupiter. As it descended through 97 miles of the Jovian atmosphere, it collected 58 minutes of data about the weather. The probe was eventually destroyed as it continued to descend through the hydrogen layer beneath the Jovian cloud tops. It's amazing that the probe was able to survive as long as it did in Jupiter's harsh and dangerous environment.

*Galileo's* Probe.

When *Galileo* finished its mission on Jupiter, scientists programmed it to crash into Jupiter. It disappeared beneath the foggy surface and into the unknown.

So far, 8 NASA spacecraft have collected information about Jupiter. Another spacecraft, the *Juno*, will arrive at Jupiter, allowing NASA to continue studying the planet.

**THE JOVIAN TIMES**

**Galileo News**

Almost everything we know about Jupiter comes from the unmanned spacecraft called Galileo. In 1989, Galileo began an interesting journey to Jupiter, taking 6 years to get to there. It would not normally take that long to get to Jupiter, but this spaceship took the long way around. When Galileo left Earth, it traveled away from Jupiter, toward Venus. Why?

**Moon Tune**

Jupiter is the second brightest planet in our solar system. Jupiter is so large that you can see some of its details, even its moons, with just a pair of binoculars or a small telescope! The way to see the moons with binoculars is to steady the binoculars against something solid like a car or a mailbox.

**Got Gas?**

If you look at the picture on page 112, you will see a big spot on Jupiter. That's the Great Red Spot, a giant storm (like a hurricane) that travels around the planet. This storm is many times stronger than the worst hurricane on Earth, and it has been raging on Jupiter for more than 300 years! How do we know this? Astronomers have seen it in their telescopes since at least 1665. The Great Red Spot is twice the size of Earth! That is one huge, long-lasting storm.

In pictures, we see beautiful stripes crossing Jupiter. However, those stripes aren't all that beautiful when you get close. Those stripes are clouds of violent storms. Jupiter has lightning that is much more powerful than the lightning on Earth.

**Jupiter: An Overprotective Parent**

Jupiter is like a protective mother to Earth. You see, Jupiter is so massive that it has a very strong gravitational pull. Do you remember we talked about the rocks flying around in space? Meteoroids hit Earth and burn up, but comets and asteroids are much, much

# Activity 9.2

## Create a Newspaper

You are going to design the front page of a newspaper all about Jupiter. Get a newspaper and study the front page. An example is shown to the left. Try to make your layout like a real newspaper. You might want to begin with a **sketch** (a pencil drawing) of what you want your newspaper to look like. The name of the newspaper should go across the top. Name your newspaper something catchy. After you have written the name of your newspaper across the top, put the date in smaller writing below the title. Below the title, there should be a variety of different headlines. The **headlines** are the bigger letters that tell you what the story below is going to be about. Newspaper reporters write interesting headlines that make people want to read the story. Write exciting headlines about Jupiter: the storms, its moons, or the *Galileo*. The story below the headline is called an **article**. Write a short article below each of your headlines. Put your finished paper in your notebooking journal.

## What Do You Remember?

How does Jupiter protect our planet? Why is Jupiter like a little sun? What is the Great Red Spot on Jupiter? Why does Jupiter have stripes? Name Jupiter's 4 largest moons. Why are they called Galilean moons? Can you describe Amalthea? What do you remember about the spacecraft, *Galileo*? What was your favorite part of this lesson?

LESSON 10

# SATURN

## *wisdom from above*

The astronomer Galileo was the first person to see Saturn's rings. Using a telescope, he spotted them in 1610. That's more than 400 years ago! Ever since that day, scientists have been studying Saturn's rings.

*The LORD by wisdom founded the earth,*
*By understanding He established the heavens.*

Proverbs 3:19

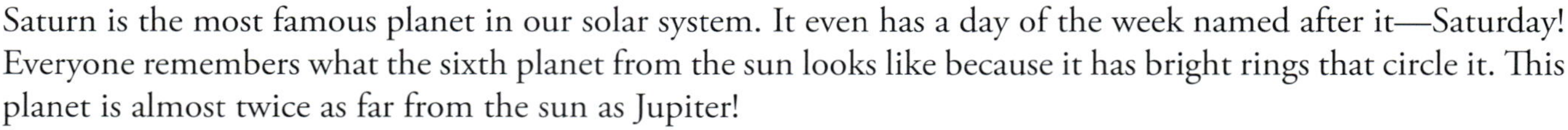

Saturn is the most famous planet in our solar system. It even has a day of the week named after it—Saturday! Everyone remembers what the sixth planet from the sun looks like because it has bright rings that circle it. This planet is almost twice as far from the sun as Jupiter!

Saturn is easy to see in the sky, and has been studied throughout history. However, the rings around Saturn were not discovered until 1610, when Galileo looked at Saturn through his telescope. Because of the weakness of Galileo's telescope, the rings looked like two little handles sticking out on either side of Saturn. Thinking they were cup handles or odd-looking moons, he was very confused by them. Forty-five years later, another scientist named Christian Huygens (hi' ginz) used a more powerful telescope to study Saturn, and he realized that they were rings!

This picture shows the view of Saturn from Earth during different times in Saturn's orbit around the sun.

Saturn is tilted as it rotates. In fact, its tilt is similar to Earth's tilt. Do you remember how this tilt affects Earth? It gives Earth its seasons. Since Saturn is tilted like Earth, it also has seasons. Saturn's tilt causes something else as well. It causes the view of Saturn from Earth to change. During part of its trip around the sun, we see mostly the edges of Saturn's rings. As it continues in its orbit, however, our view of Saturn changes, and its rings are much more visible.

Saturn is a little smaller than Jupiter, making it the second biggest planet in our solar system. Saturn and Jupiter together are the largest gas giants. The

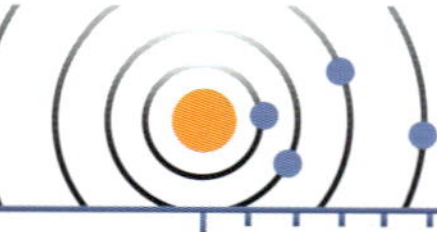

Saturn is 886 million miles away from the sun.

next two planets further out, Uranus and Neptune, are also gas giants, but Jupiter and Saturn are much larger. To see the difference in size between Earth and Saturn, get a marble and a soccer ball. The soccer ball would be Saturn, and the marble Earth. It would take 763 Earths to fill up Saturn!

## Twins

God made Saturn a lot like Jupiter. They are almost like twins, and astronomers often call them twin planets. Do you remember the difference between a terrestrial planet like Earth and a gaseous planet like Jupiter? Like Jupiter, Saturn is a gaseous planet mostly made of hydrogen and helium.

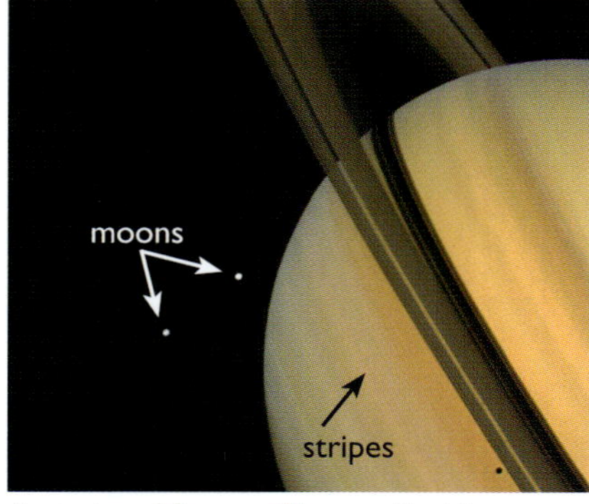

A photo of Saturn. The small white spots are two of Saturn's moons.

This is an artist's painting showing clouds of icy ammonia across Saturn's upper atmosphere.

Being further away from the sun, Saturn is much colder than Jupiter. Saturn is usually about 300 °F below zero in its upper atmosphere. The very center of Saturn, however, is hot, hot, hot—with a temperature greater than 20,000 °F! Like Jupiter, Saturn puts out its own heat, rather than getting all of its energy from the sun.

Saturn is beautiful, but its beauty is deceiving for Saturn is a horridly cold and stormy planet. Although Saturn's storms are not quite as violent and long-lasting as Jupiter's storms, you wouldn't want to be caught in one. Do you see the stripes that go around the planet? Those are bands of clouds, and the storms occur within those clouds. The stripes add to Saturn's beauty, but the winds within those stripes blow fiercely at about 1,000 miles per hour. That's 5 to 10 times stronger than the strongest hurricane on Earth! These winds get their energy from the heat rising from Saturn's interior. We couldn't survive if Earth had winds that strong blowing on it. That's why God created Earth differently than the other planets.

As the gases in Saturn's interior warm up, they rise until they reach a level where the atmosphere's temperature is cold enough to freeze them into particles of solid ice. Icy ammonia forms the outermost layer of clouds. The clouds look yellow because ammonia reflects the sunlight.

Saturn is the only planet in our whole solar system that is less dense than water. If something is denser than water, it will sink. If something is less dense than water, it will float. So, if there were a body of water large enough, Jupiter would sink and Saturn would float. In fact, Saturn is the only planet that would float. That would be one giant bath toy!

# Activity 10.1

## Draw a Venn Diagram

A Venn diagram is a wonderful tool that helps you compare and contrast different things. The drawing on the next page is an example of what a Venn diagram looks like. Use a sheet of paper to make your own Venn diagram to compare and contrast Jupiter and Saturn. Study your notes from Jupiter (or read the Jupiter lesson

**(Continued on next page.)**

again) to remember what you learned. Write the things that are the same about both planets in the overlapping part of the diagram, and then record those things that are different in the ovals outside of the overlapping part. Place your completed diagram in your notebooking journal.



**Take a moment to tell someone about Saturn. You can use your Venn diagram to show them about both Jupiter and Saturn.**

## Ring System

There are thousands of rings around Saturn. Yes, thousands! It looks like one or two rings when you look at Saturn through a telescope on Earth, but spacecraft have gotten close enough to see that there are many more than that. The rings are made of dust, ice, and rocks that orbit around the planet. Some of the rocks are the size of a little pebble, and some are the size of a whole house! If you could stand on Saturn, you would see giant chunks of rock and ice, sometimes as large as a house, swirling above you. Astronomers aren't completely sure what formed these rings. Some think that giant asteroids or comets crashed into one or more of Saturn's moons, smashing them to pieces and creating the rings. Others think that the rings were once asteroids or comets themselves, but they got so close to Saturn that broke into tiny pieces encircling the planet. Saturn's rings are one more neat mystery that God has given us. Perhaps you will one day figure out the mystery of the rings!

The colors are false in this photo of Saturn's rings. They are used to show the large number of rings.

Saturn's rings also contain whole moons. Do you know what a shepherd does? A shepherd takes care of sheep, keeping them all together and leading them where they should go. Saturn has **shepherd moons**, called that because they herd the rings as if they were sheep, and keep them from spreading out too far. The

An artist's idea of what Saturn's rings look like.
*Digital artwork by Dr. David Heatley*

two best-known shepherd moons are **Pandora** (pan dor' uh) and **Prometheus** (pruh me' thee us).

Saturn's rings are so thin that sometimes you cannot see them even with a telescope. Remember, the view of Saturn from the Earth changes, and if the tilt is just right, we can only see the edges of the rings. The rings are less than one mile thick, which is tiny for a gigantic planet like Saturn. That makes them hard to see in some telescopes. Although they are thin, they are very wide. Because they are wide, when Saturn's tilt is right, we can see the rings quite well with a telescope.

This photo shows the thin rings around Saturn.

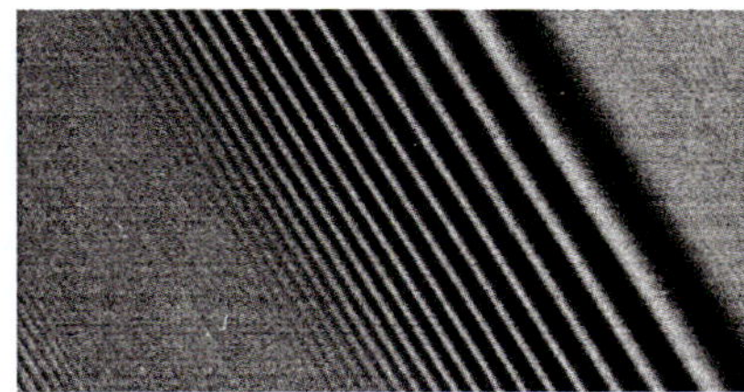
A close-up of Saturn's rings taken by the Cassini spacecraft on July 1, 2004. The rings look like a collection of rock and ice particles.

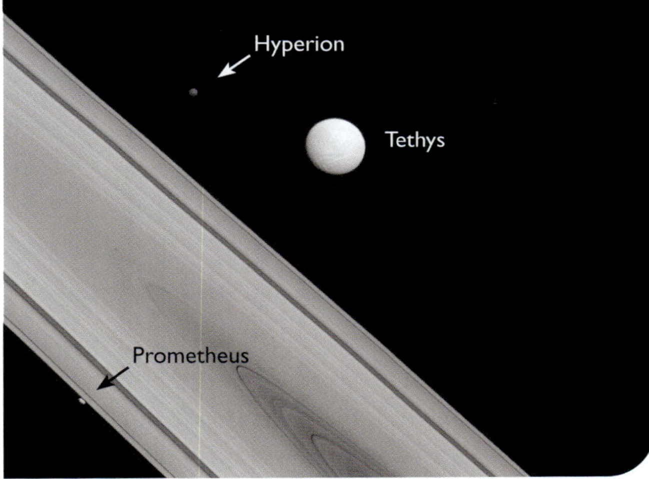

The Prometheus moon (63 miles across) is visible bottom left in the image. The Hyperion moon (168 miles across) appears above left, and the Tethys moon (168 miles across) is the largest one seen in the image.

## *think about this*

Saturn's rings look like a wide, flat surface, almost as wide as the distance between Earth and the Moon. But you know the truth is that they are made of millions of pieces of ice and rock. Scientists say that the rings may be the solar system's largest traffic jam!

**What do you remember about Saturn so far?
Explain in your own words all that you have learned.**

## Fast Rotation

Have you ever spun a top? A top twirls extremely fast. Saturn is spinning around so fast that it's daytime for just over 5 hours, and then it's nighttime for just over 5 hours. Guess what would happen if you went to bed on Saturn and slept for 10 hours. When it was about time to wake up, it would just be getting dark again! Around lunch, it would begin getting light again. When you were ready to go back to bed, it would get dark again. Saturn's fast spin also makes the planet bulge at the middle and flatten at the top. This makes Saturn look like somebody is squashing it. Do you remember that Jupiter rotates quickly, too? Do you remember that Jupiter's fast rotation is one reason it has such strong winds and terrible storms? The same is true for Saturn. The fast rotation of these two gas giants is just one more reason that they are considered twin planets.

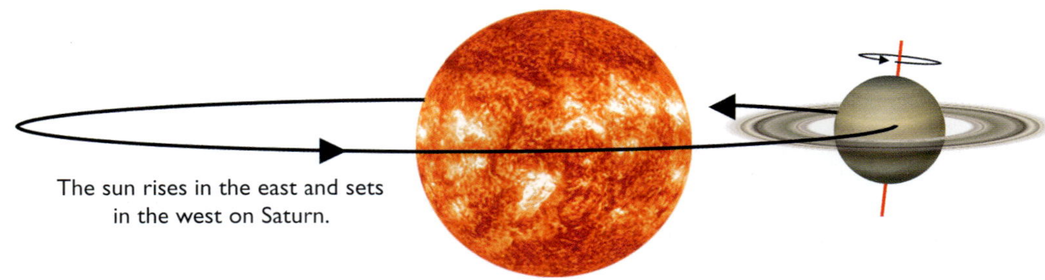

The sun rises in the east and sets in the west on Saturn.

Saturn is even farther from the sun than Jupiter. It takes almost 30 Earth years for Saturn to orbit the sun just one time. If you were 2 Saturnian years old, you would be 60 years old on Earth. You might even be a grandparent!

## Who Named Saturn

Saturn is named after the Roman god Saturnus.

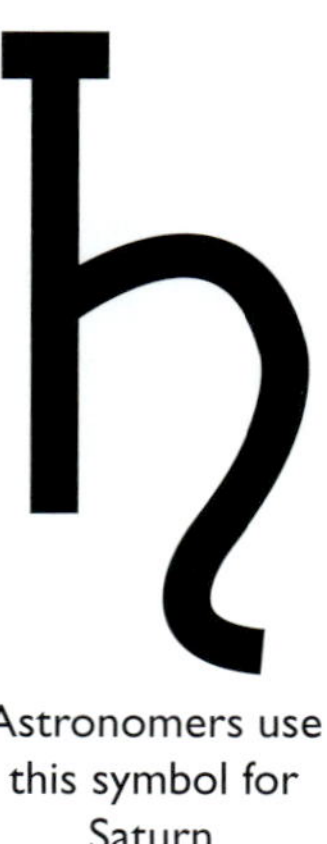

Astronomers use this symbol for Saturn.

## Finding Saturn in the Night Sky

Saturn is easy to see in the night sky. It isn't as bright as Jupiter, but you can tell it apart from the stars because the light is steady and not twinkling. If you have a telescope, you can even see the rings of Saturn. During certain years, Saturn is tilted on its side so that you can see the rings perfectly. If you go to **www.apologia.com/bookextras**, you will find links to websites that will help you find Saturn.

## Cassini Mission

An artist's idea of what *Cassini* looks like orbiting Saturn.

When a spacecraft travels somewhere, it is on a *mission*. The mission of the spacecraft exploring Saturn and its moons is called the **Cassini** (kuh see' nee) **Mission**. *Cassini* is a very big spacecraft. It is taller than a two-story building, and weighs more than 6 tons. In other words, it's about the weight of an empty 30-passenger school bus. It took a lot of rocket power (and money) to launch *Cassini* into space. As *Cassini* orbits Saturn, it takes pictures of Saturn and its moons. It also sent a probe named **Huygens** to collect and test samples of dirt from one of Saturn's moons called Titan. A link found at **www.apologia.com/bookextras** will enable you to follow the progress of the Cassini mission.

# Saturn's Moons

If you spent the night on Saturn, it wouldn't be very dark because astronomers have discovered 53 moons that orbit Saturn! Only Jupiter has more moons than Saturn.

Saturn and its 7 largest moons. The relative sizes of the moons are not correct.

Saturn's biggest moon is named **Titan** (tie' tun). Titan is bigger than the planet Mercury. It is the only satellite in the solar system known to have an atmosphere. This atmosphere, consisting of clouds thick with nitrogen and methane, makes Titan look orange. These clouds also make it difficult to see the surface of the moon from a spaceship. The 7 largest moons of Saturn are shown in the picture above. Since many of Saturn's moons are small, irregularly-shaped rocks rather than nice round moons, they are likely space rocks that got caught in Saturn's gravitational pull.

In 2005, the *Cassini* spacecraft launched a probe called *Huygens* to Titan. The surface was soft (like clay) where the probe landed and images taken after landing showed several small stones and pebbles. *Huygens* discovered that the temperature of Titan averages -290 °F. The winds there were really fast—as high as 270 miles per hour! Lightning was also detected on Titan.

Saturn's Titan moon.

Artist interpretation of Titan.

A photograph of Titan's surface taken from the Cassini Huygens probe as it descended to the moon's surface.

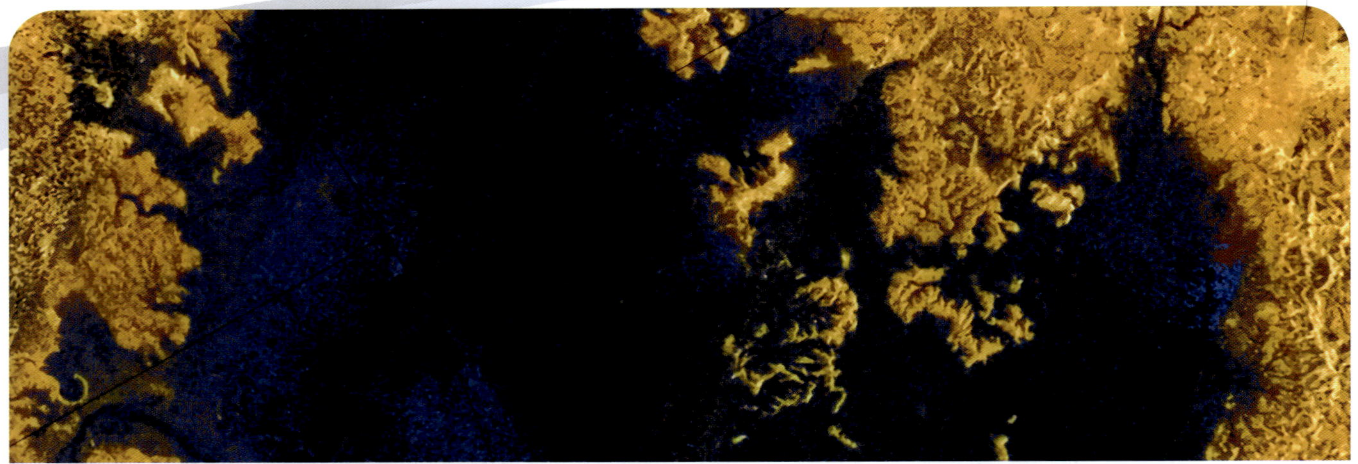

Titan's Ligeia Mare, a hydrocarbon sea.

*Cassini* captured this image of Titan; the darker spots are some of the largest seas and some of the many hydrocarbon lakes that are present on Titan's surface. Titan is the only place in the solar system, other than Earth, that has stable liquids on its surface. In this case, the liquid consists of ethane and methane rather than water. Titan's largest sea is Kraken Mare.

**Take a moment to explain the Cassini mission and Saturn's moons.**

# Activity 10.2

## Launch a Rocket

*Cassini* was sent into space with the help of a rocket weighing about 1,038 tons. Rockets launch spacecraft, such as *Cassini*, out of our atmosphere and into space using the force that results from a chemical reaction. You are going to explore the power of chemical reactions by launching your own rocket.

### You will need:

- Adult supervision
- Empty water bottle
- Eye protection (such as safety goggles or glasses)
- 2 or more Alka Seltzer® tablets
- Water
- Clay plug
- Paper towel
- Imagination

### You will do:

1. Use your imagination to decorate your empty and open water bottle so that it looks like a rocket. Be sure to keep the opening free and at the bottom of your rocket.

**(Continued on next page.)**

2. Depending on the size of your water bottle, crush one (small bottle) or two (regular bottle) Alka Seltzer tablets.
3. Tear about a 4X4 inch square from your paper towel and roll your crushed tablets up tightly into the paper towel so that the roll can fit through the mouth of your water bottle. Set it aside.
4. Use clay to form a tight cork-like seal for the mouth of your water bottle. **DO NOT use the cap that came with your bottle!** Set your cork aside.
5. Gather your rocket, fuel (rolled up tablets), and cork. Go outside and find a clear spot where your rocket can stand up easily AND everyone can move away quickly.
6. Fill your "rocket" about 1/3 full with water. Insert your paper towel fuel rod and quickly put your cork in place.
7. Set your rocket on the ground and move away quickly.
8. **Your rocket will launch with a very powerful force!** Enjoy the reaction.

## Discussion

Your rocket launches because there is a chemical reaction when the Alka-Seltzer® mixes with the water. This reaction causes gases to build up in the container. These gases are very active, moving around a lot. As the reaction continues, more and more gases develop. These gases need more room in which to move around, so they begin to expand inside the container. They push harder and harder until they push the cork out of the container. When they do this, the gases fly out the open end of the container, causing the whole rocket to go flying into the air.

The Centaur rocket that launched *Cassini* into space. Notice how small the engineers look!

*Cassini* was launched at night.

# What Do You Remember?

What is Saturn made of? Why would Saturn be an unpleasant place to visit? Which planet is considered Saturn's twin? What are Saturn's rings made of? What do shepherd moons do? How many years does it take Saturn to orbit the sun? Why does Saturn look as if it is being squeezed? What is the name of the spacecraft mission which is currently studying Saturn? What was your favorite part of this lesson?

LESSON 11

# URANUS

## *wisdom from above*

Astronomers spend their days peering into the heavens. When Uranus was discovered, William Herschel said, "I have looked further into space than any human being did before me" (Gibson 1913, 211). Imagine having that privilege!

Beyond the 2 enormous yellowish-gold planets of Jupiter and Saturn is another gas giant called Uranus (yur' uh nuhs). Uranus is blue-green in color and is also the third largest planet in the solar system after Jupiter and Saturn.

What makes Uranus appear blue-green? It's the atmosphere, of course! Isn't that always the case with the gas giants? Uranus has an atmosphere made of helium and hydrogen, just like Saturn and Jupiter, but it also has another gas in its atmosphere called **methane** (meth' ayne). Methane absorbs red light and reflects blue light, so we see the planet as blue in color. Humans cannot breathe methane, so I guess we won't be vacationing on this planet either.

Uranus and its rings. Although it looks like there only are only 2 or 3 rings, there are actually a total of 13.

Uranus also has 2 sets of rings. The inner ring system has 11 rings. The outer ring system is made up of 2 brightly colored rings discovered by the Hubble space telescope in 2003. The rings orbit Uranus vertically. That means circle the planet up and down, rather than around the middle like the rings of Saturn.

As you can imagine, Uranus is freezing cold. It's about 350 °F below zero. That's 50 °F colder than Saturn!

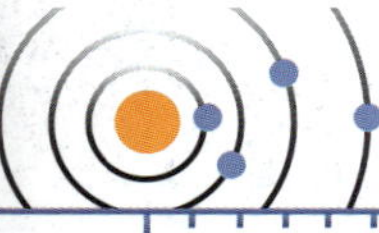

Uranus is 1,784 million miles away from the sun.

## Eureka!

**Eureka** (yuh ree' kuh) means, "I have found it." Exciting discoveries are often called *eureka moments.* One such eureka moment was the discovery of Uranus in 1781. It was exciting because it was the first planet to be discovered in a long, long time! Mercury, Venus, Mars, Jupiter, and Saturn were well known throughout history. Uranus was something completely new.

This is a photo of Mimas, a moon of Saturn that was discovered by William Herschel. The huge crater is called Herschel crater.

William Herschel and his sister Caroline accidentally discovered Uranus. Their dad was a musician who taught them his favorite subjects: music, math, and astronomy. They lived and were homeschooled in Germany, and grew up to be very important musicians in England. Before they discovered Uranus, William and Caroline earned a living by singing and playing music in the opera. But, in his spare time, William built telescopes. They were *amateur* (beginner) astronomers, just as you are!

One day, while looking through one of their homemade telescopes, William and Caroline noticed a light that looked like a disk, not a star. At first, William and Caroline, thought it must be a comet. As they continued to watch this disk for months, they noticed that its orbit around the sun was nearly circular. Remember, comets have very elliptical orbits, while planets have orbits that are nearly circular. This fact, along with Galileo's discovery hundreds of years before that planets look like disks, led them to determine that it was a planet, not a comet!

Once William and Caroline discovered Uranus, they became famous. They left their musical careers when William was appointed Court Astronomer by King George. Caroline was his paid assistant and the first woman to be paid for scientific study. Caroline discovered many comets in her work as an astronomer, and William discovered many things, like 2 of Saturn's moons. Later astronomers honored their work by naming several craters in the solar system, including one on our Moon, one on Mars, and one on Saturn's moon, Mimas, after the Herschels.

### *think about this*

Guess what happens when you find a planet? You get to name it! Since William Herschel was a devout Christian, he didn't want to name Uranus after a false Roman god. Instead, he wanted to name Uranus after his patron, King George III. Not everyone loved King George, however. Many people wanted Herschel to name the new planet after himself. In the end, scientists kept with tradition and named Uranus after another Roman god.

# Activity 11.1

## Create a Play About the Discovery of Uranus

You are going to write a short play about the discovery of Uranus. Older students will write it out; younger students will dictate their ideas and words to their parent/teacher to write out. Everyone should act it out! I will remind you of the facts mentioned in this chapter, and I will add a few more facts, in case you want to make your play even more interesting.

**William and Caroline Herschel were brother and sister, living in Germany. Their parents had six children. Their father was a musician and taught all of his children music, as well as mathematics and astronomy.**
**Suggestion:** You could do a scene with them at home learning music, astronomy, and math.

**William and Caroline moved to England. This made their mother mad because she wanted Caroline to be her servant and housekeeper for the rest of her life.**
**Suggestion:** You could do a scene with their mother begging them to stay so Caroline could be her servant.

**In Bath, England, they became very important musicians in the opera. William played the piano and wrote the plays for the opera, and Caroline was a singer. In William's spare time, he built telescopes, continually making them more and more powerful in order to see deeper and deeper into space. People in the town loved and bought his great telescopes.**
**Suggestion:** You could do a scene with them in the opera or making telescopes and selling them.

**One day, William and Caroline noticed something in the sky that didn't look like a normal star. It looked like a disk. They thought it must be a comet. They watched it and watched it and debated about it. Finally, its nearly circular orbit made them realize it was a planet!**
**Suggestion:** This is the most important scene to have in your play.

**After this great discovery, the government paid them to be full-time astronomers and telescope builders. They discovered many more comets and other important things.**
**Suggestion:** You can end with the government making them full-time astronomers.

(Continued on next page.)

## You will do:

1. Decide how many people will be in your play. You may decide to have only William and Caroline. If you have lots of friends or family who would enjoy acting in your play with you, you could include William and Caroline's parents, brothers and sister, the government officials, other singers in the opera, and customers buying telescopes.
2. Make a list of characters and their names.
3. Decide if this will be a short play with one or two scenes, or a longer play with up to five scenes.
4. Decide how the play will end:
   - Will it end with the discovery of Uranus?
   - Will it end with the government paying them?
   - Will it end with them being important astronomers, discovering many other things?
5. Decide where the first scene will take place and write what will happen in that scene.
   - Will it be at home in Germany with Dad teaching everyone math?
   - Will it be with them departing for England, thanking their dad for teaching them so well?
   - Will it be with them performing at the opera or building and selling telescopes?
   - Will you only have one scene in which they are looking through the telescope on the day they discover that Uranus is a planet?
6. Think about what people will say to one another in this scene and write it out. In plays, the name of the person speaking is written first with a colon after it, like this:
   **Caroline:** I think it's a comet.
   **William:** But it looks like a disk! Comets look like smudges.
   **Caroline:** What else could it be? Surely it is not another planet!
   **William:** Why, I never thought of that.
7. Repeat steps 5 and 6 for each scene.

## Discussion

This activity will be even more fun to act out if you use *props*, such as telescopes. Performing your play will help you remember the very important story of the discovery of Uranus.

# Orbit and Rotation

We might call Uranus our sleepy planet because it is lying down. It rotates sideways, not up and down like other planets. God may have created Uranus as a tipped over planet or it might have toppled over onto its side when a giant comet hit it. What do you think?

Because Uranus is so far away from the sun, it takes a long time, almost 84 Earth years, for it to make one revolution around the sun. Have you noticed that the farther a planet is from the sun, the longer it takes for it to go one time around the sun? For nearly 20 Earth years during each orbit, the sun shines directly over each pole for a long summer, while the other half has a very cold and dark winter. This happens because of the way Uranus rotates. Since it rotates on its side, Uranus always faces in the same direction. You can see this in the image. Notice how the northern hemisphere always is to the right of the page while the southern hemisphere always is to the left.

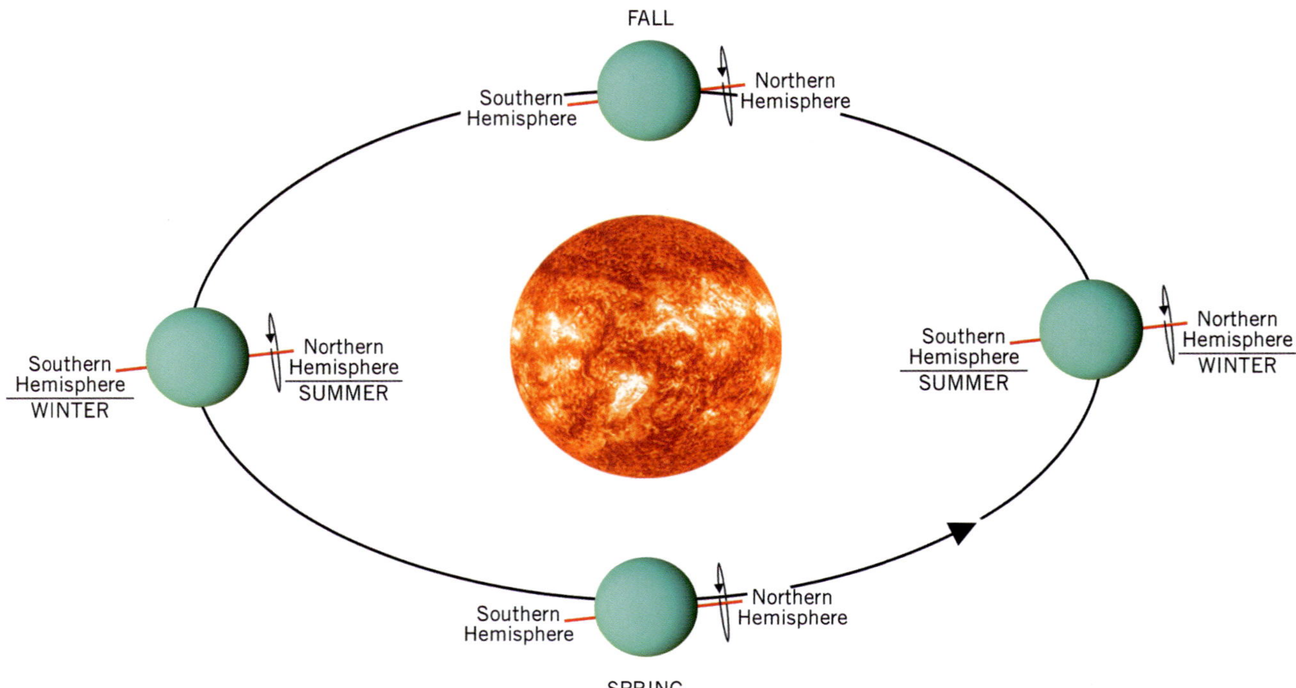


Do you know that Uranus is also spinning in the opposite direction as compared to most of the other planets. This means the sun rises in the west and sets in the east. Can you think of another planet that spins opposite of Earth? It's Venus. You learned about its strange rotation in lesson 4. Although each year on Uranus is extremely long, each day is just over 17 hours, which is not too different from our 24-hour day on Earth.

**What is unique about the orbit of Uranus?**

# Moons

Can you guess how many moons Uranus has? It has 27! At least 27 that we know about. Uranus used to be known as the planet with the most moons, until astronomers found even more moons orbiting both Jupiter and Saturn. Most of the names of planets and moons come from Roman myths and legends, but not the moons of Uranus. These moons are all named after characters in the writings of either William Shakespeare or Alexander Pope. Let's take a peek at some of these moons.

This is a composite photo of five of Uranus' moons. The sizes are not correct because the artist wanted to make it look like you are closest to Ariel.

This is a photo of Titania taken by *Voyager 2*. Notice the large canyon in the bottom right part of the picture.

This is a photo of Oberon taken by *Voyager 2*. There are many craters on the surface of this moon.

## Titania

**Titania** (tih tan' yuh) is Uranus's largest moon. Pictures taken by the *Voyager 2* spacecraft show that Titania is made of ice and rock. The pictures also show that the surface of Titania has enormous canyons. One of the largest canyons is about 1,000 miles long, which is roughly the diameter (distance across) of Titania. That's about half the size of Earth's Moon.

## Oberon

**Oberon** (oh' buh rahn) is the second largest moon of Uranus. Even though it was discovered over 200 years ago, in 1787, we didn't know much about this moon until recently. When *Voyager 2* passed it during a flyby, pictures showed that it has many craters and is made from ice and rock. It also has at least one large mountain that rises 4 miles above the surface. Oberon is about 950 miles across—about one fourth the size of Earth's Moon.

## Cordelia and Ophelia

Both Cordelia and Ophelia are believed to be made from ice and rock. Cordelia is the closest moon to Uranus.

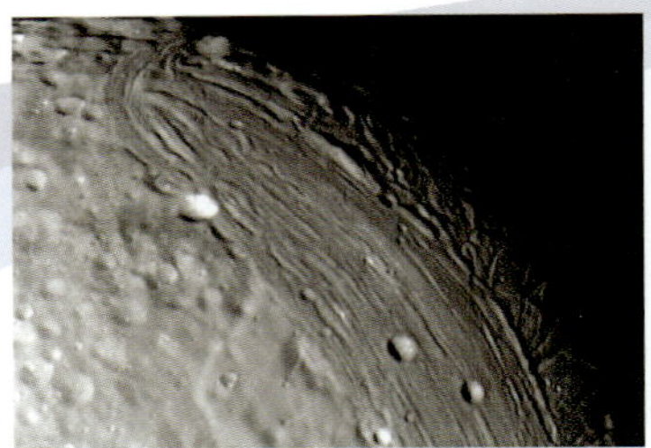

This is a photo of Miranda's surface taken by *Voyager 2* on January 24, 1986.

## Miranda

**Miranda** (mi ran' da) is a pretty small moon—about 300 miles across. It's also a pretty strange moon. It looks like it was pieced together from parts that don't fit making the surface very rough. It's covered with ridges and valleys that are separated from the more heavily cratered surface areas. Its canyons are extremely deep—some are 12 times as deep as the Grand Canyon.

Miranda, being so small, doesn't have a very strong gravitational pull. In fact, if a rock dropped off the edge of its highest cliff, it would take 10 minutes for it to reach the bottom! That's a slow fall.

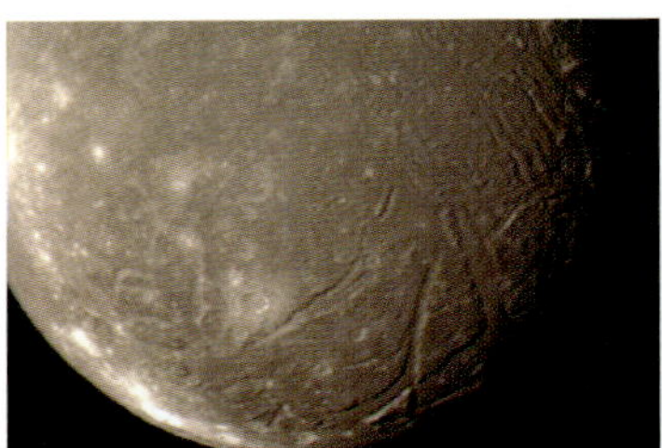

A photo of Ariel taken by the *Voyager 2* spacecraft.

## Ariel

About 720 miles across, **Ariel** (air' ee al) is one of the 5 largest moons orbiting Uranus. We might say Ariel is shiny because it has the brightest surface of the 5 big moons. It's a big bright moon made of ice and rock. Like our Moon, it keeps the same side facing toward Uranus as it orbits the planet. Ariel orbits Uranus every 2.5 Earth days.

## Umbriel

**Umbriel** (um' bree al) is the third biggest of Uranus's moons. Can you guess what it's made of? Pretty much the same stuff as the others—ice and rocks. It is about 750 miles in diameter, the same size as Ariel, but only reflects half as much light. So, it's a lot darker. We don't think it has an atmosphere or a magnetic field. This dark moon orbits Uranus every 4.2 Earth days.

A photo of Umbriel taken by the *Voyager 2* spacecraft.

**Can you use the illustrations here to tell someone about Uranus's moons?**

## Who Named Uranus

You know that Herschel discovered Uranus, but it was a German astronomer named Johann Elert Bode who named the planet after an ancient Greek god of the sky. His reasoning was that since Saturn was the "father" of Jupiter, the new planet should be named for the "father" of Saturn.

Johann Elert Bode

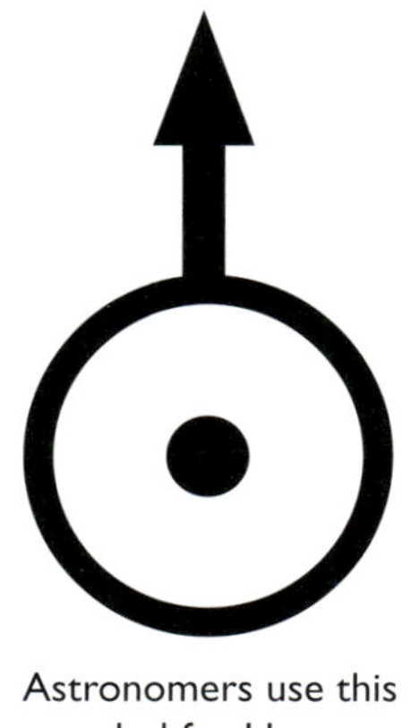

Astronomers use this symbol for Uranus.

# Activity 11.2
## Make Clouds

### You will need:
- Adult supervision
- Glass jar
- Ice
- Match
- Large Ziploc® bag
- Very hot water

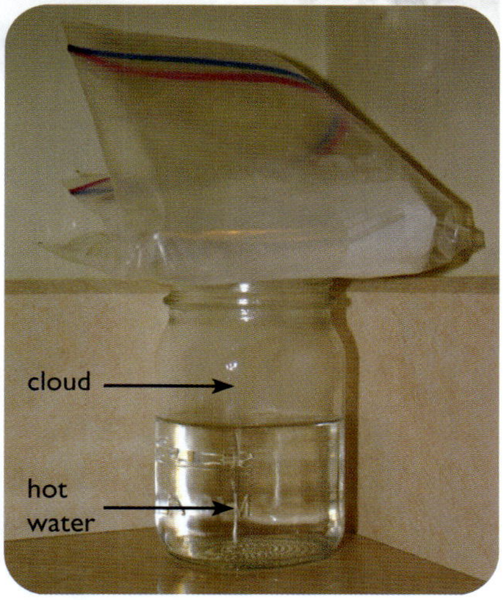

### You will do:
1. If the jar has a lid, remove it.
2. Turn on the hot water and let it run so that it gets really hot. To make this work even better, have your parent/teacher put water in a coffee mug and heat it up in a microwave oven.
3. As the water is heating up, fill the bag with ice and zip it closed.
4. Fill the jar halfway with the hot water. **Be careful! Don't burn yourself!**
5. Have your parent/teacher strike a match and drop it into the water so that it goes out when it hits the water.
6. Place the bag of ice over the top of the jar.
7. Watch as a cloud forms in the jar.

### Discussion
The cloud forms in the jar because the hot water evaporates, turning into a gas, which we call **water vapor**. The water vapor rises, but it then cools when it hits the colder air that has been cooled down by the ice. This cooling causes the water vapor to become a liquid again, creating a cloud. The smoke from the match supplies some particles in the air above the water. This makes it easier for the water to turn into a liquid when it cools.

This is what happens on the gas giants. The inner core of each planet is very hot, while the gases that make up its atmosphere are extremely cold. Gases rise from the hot core. When they cool off in the icy-cold atmosphere, they turn into liquids and then solids (like ice crystals), making clouds. Now remember, the clouds on the gas giants are not made of water. They are made of methane and other chemicals. Still, the overall process is very similar to what you saw here.

## What Do You Remember?

What chemical makes Uranus blue-green in appearance? Why was it so exciting to discover Uranus? Who discovered Uranus? How were they educated? How long does it take Uranus to orbit the sun? How many moons does Uranus have? What is Uranus's largest moon? The sun rises in the west and sets in the east on 2 planets; which 2 planets are they? What was your favorite part of this lesson?

LESSON 12

# NEPTUNE

## wisdom from above

Scientists typically build on the knowledge that came before them. In discovering Neptune, astronomers used, "...the irregularities of the motion of Uranus. Their amazingly precise observations and calculations predicted the existence of a then-unknown planet" (Schaub and Junkins 2003, 489) There is always more to learn about creation.

*"I saw every work of God, I concluded that man cannot discover the work which has been done under the sun. Even though man should seek laboriously, he will not discover; and though the wise man should say, "I know," he cannot discover."*

Ecclesiastes 8:17

Great Dark Spot

Neptune's color is a beautiful, bright blue.

# Neptune

Beyond Uranus is the last planet in our solar system. Its name is Neptune (nep' toon) and it has a bright blue color. Although Neptune is the fourth-largest planet in the solar system, it is so far from the sun that you need a telescope to see it.

What makes Neptune appear blue? As with any gas giant, it's the atmosphere. In addition to helium and hydrogen, Neptune has **methane** (meth' ayne) in its atmosphere. Methane absorbs red light and reflects blue light, so we see this planet as blue. Neptune is a deeper blue than Uranus because it has more methane in its atmosphere.

When Genesis tells us that God declared all things *good*, it means that He takes great pleasure in His gorgeous creation. The Bible also tells us that the highest heavens belong to God. It must have given Him joy to create such amazingly beautiful spectacles in the heavens. They are His. Of course, God also knew that we would one day be able to see the outer planets and enjoy them as well. As the Bible says, *"The heavens are telling of the glory of God; And their expanse is declaring the work of His hands" (Psalm 19:1).*

Neptune's rings as photographed by *Voyager 2*.

As you can imagine, Neptune is freezing cold. Its atmosphere is about 353 °F below zero. That's about the same temperature as Uranus, and 65 °F colder than Saturn.

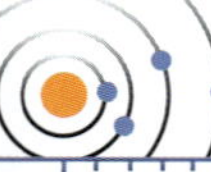

Neptune is 2,795 million miles away from the sun.

Neptune also has 6 rings that we know about so far. They are hard to see unless a picture is taken at the correct angle with light from the sun. The 2 main rings, shown in the picture are about 33,000 miles and 39,000 miles from Neptune. Neptune's rings contain dust-sized particles making them different from most of Uranus's and Saturn's rings which are made from rocks and ice.

# Activity 12.1
## Make Ice Cream!

Neptune is a very cold planet, and everything on it is frozen. To freeze something, you must cool it down. How much do you have to cool it down? That depends. Different chemicals have different freezing points. For example, water freezes at 32 °F, so we call 32 °F the **freezing point** of water. Other chemicals have other freezing points. If you mix other chemicals with water, the freezing point of the mixture is lower than the freezing point of water. We are going to do an experiment to see how this works. The amounts listed make one small serving. You can increase the amounts to make more.

### You will need:
- 2 tablespoons powdered sugar
- ½ cup whipping cream (whole milk or half-and-half will work)
- ¼ teaspoon vanilla
- 6 tablespoons rock salt
- 1 pint-size Ziploc® plastic bag
- 1 gallon-size Ziploc® plastic bag
- Several ice cubes

### You will do:
1. Put the milk, vanilla, and sugar into the small bag, and seal it.
2. Fill the large bag half full of ice, and add the rock salt.
3. Place the small bag inside the large bag, and seal the large bag.
4. Shake vigorously until the mixture in the small bag is ice cream (about 6-8 minutes).
5. Wipe the salty water and ice off of the small bag.
6. Open the bag and pour its contents into a bowl.
7. Enjoy the results of your experiment!

### Discussion

Notice that the ice cream you made was frozen solid. However, the salt water in the outer bag was not frozen. Why did the mixture in the inner bag freeze when the mixture in the outer bag did not? When the ice and salt mixed, the salt did two things. First, it lowered the freezing point of water, making the ice melt. Second, it cooled off the melting ice, making the mixture colder. Usually, if water gets to a temperature of 32 °F or below, it freezes. However, as the salt cooled the ice, it lowered the freezing point of the water. The mixture in the outer bag was about 28 °F once the salt and water mixed well. This was cold enough to freeze the cream in the smaller bag, making ice cream! Complex things like this happen on planets, when lots of chemical mix together.

## Eureka!

Upon finding Neptune, scientists might have once again yelled "Eureka" because they had been looking for it for quite some time! How did they know that Neptune was out there? Astronomers noticed that Uranus moved as if it were being pulled by another large object. They thought that this object must be a big planet, and began looking for it. Unlike Uranus, Neptune wasn't found accidentally. Instead, its discovery was the result of a long search.

After finding Neptune, astronomers noticed that its orbit wobbles. At first, astronomers thought that this meant it was being pulled by yet another planet. Upon finding Pluto, astronomers thought its gravity was pulling on Neptune, but they quickly realized that Pluto is too small to affect Neptune's orbit. Astronomers resumed their search for a mystery planet, but very recent work indicates that the wobbles seen in Neptune's orbit are the result of the gravity from the known planets in the solar system. Therefore, most astronomers no longer think that there is another planet in the solar system.

Astronomers use this symbol for Neptune.

## Who Named Neptune

The earliest discovery of Neptune was by Galileo, however, he mistook it for a star. In 1846, Neptune was named after the Roman god of the sea by the astronomer who discovered the planet, Urbain Le Verrier.

**Take a moment to tell someone about Neptune.**

## Orbit and Rotation

Because Neptune is farther from the sun than Uranus, it takes about twice as long for Neptune to get around the sun: 164 Earth years! Think about how long this is. If we lived on Neptune, we would never celebrate our first birthday or orbit the sun even once!

Neptune is spinning in the same direction as Earth. This means the sun rises in Neptune's east and sets in the west. Although each year on Neptune is extremely long, each day is just over 16 hours, which is not too different from our 24-hour day on Earth and very similar to Uranus's 17-hour day.

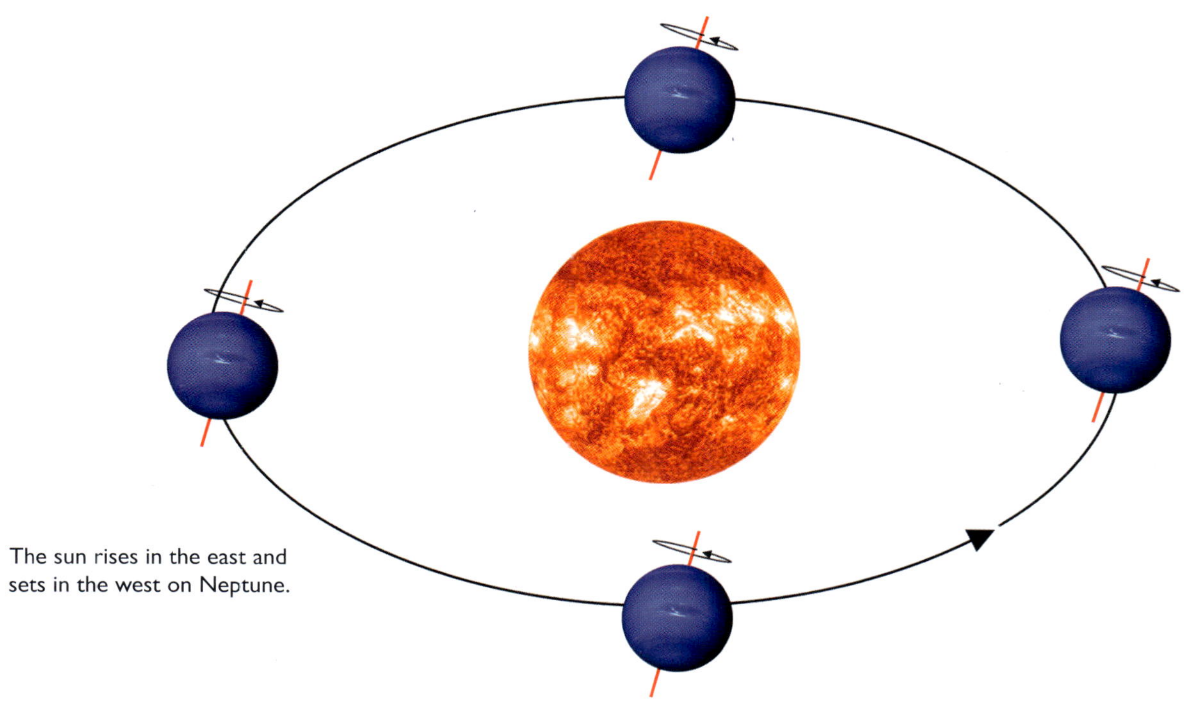

The sun rises in the east and sets in the west on Neptune.

## Atmosphere

Neptune, like Uranus, is cold and covered with methane ice. You can see a striped pattern on Neptune, just like on Jupiter and Saturn. Can you guess what the stripes are? They are clouds. Just like Jupiter and Saturn, there are storms raging on Neptune. One of these storms was called the **Great Dark Spot** and was similar to Jupiter's Great Red Spot. The Great Dark Spot was as big as the whole Earth and filled with powerful winds moving at 750 miles per hour. The Hubble Space Telescope has shown that the Great Dark Spot is now gone, and that the storm is over. However, another large storm has formed on a different part of Neptune.

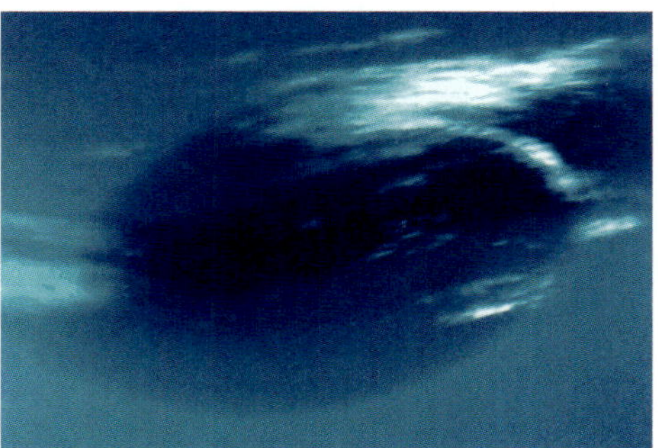

A close-up of Neptune's Great Dark Spot. Even though this storm is over, another one has appeared in Neptune's Northern Hemisphere.

Neptune is smaller than Uranus, making it the smallest of the gas giants. It has a rocky core that astronomers believe has about the same mass as Earth. That core is surrounded by liquid, which is then surrounded by the hydrogen, helium, and methane gases that make up its atmosphere. The winds on Neptune are even stronger than Saturn's winds. It would be a terribly blustery place to send a spacecraft.

## think about this

When it storms here on Earth, it rains. What is that rain made of? It is made of water. Some astronomers think that the rain in the storms of Neptune is made of diamonds. Yes, I said diamonds! Astronomers have done experiments that show that in the conditions they think exist in Neptune's storms, methane can be turned into diamond dust. Since there is methane in Neptune's atmosphere, they think that diamond dust forms in the storms and then falls from the clouds. Just imagine, diamond rain!

**Stop for a moment and tell someone a few fun facts you are learning about Neptune.**

# Moons

Astronomers have discovered 13 known moons around Neptune. The biggest is **Triton** (try' tuhn), and it's the coldest object that any spacecraft has ever visited. The average temperature on Triton is 400 °F below zero! It revolves around Neptune opposite of the direction that Neptune rotates. It is the only moon known to do that. It is also moving closer and closer to Neptune each day. These two facts make most astronomers think that Triton was not originally Neptune's moon. Instead, it was probably captured by Neptune's gravity. Triton is a fascinating moon because it is filled with **geysers** (guy' zurs). Geysers on Earth are holes in the ground that spew out hot water from deep within the planet. You can visit some of Earth's geysers at Yellowstone National Park. Triton's geysers are probably not spewing water; they are probably spewing a mixture of chemicals.

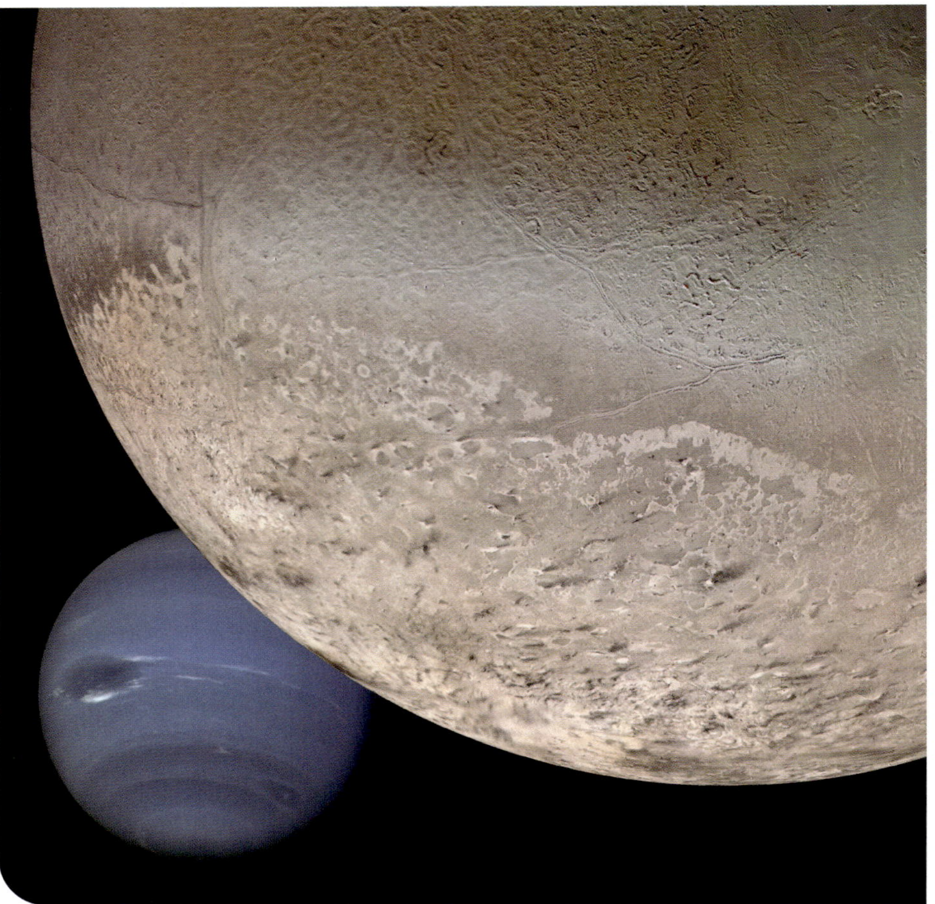

This is a photo of Neptune as seen from behind Triton. Triton is not bigger than Neptune; it appears bigger because Triton is closer in this picture.

The second largest Neptune moon is **Proteus** (proh' tee us). It was discovered in 1989 by *Voyager 2*. The temperature on Proteus is estimated to be 370 °F below zero. This moon has a diameter of 260 miles and is much smaller when compared to Earth's Moon's diameter of 2,160 miles. Proteus orbits Neptune about every 27 hours

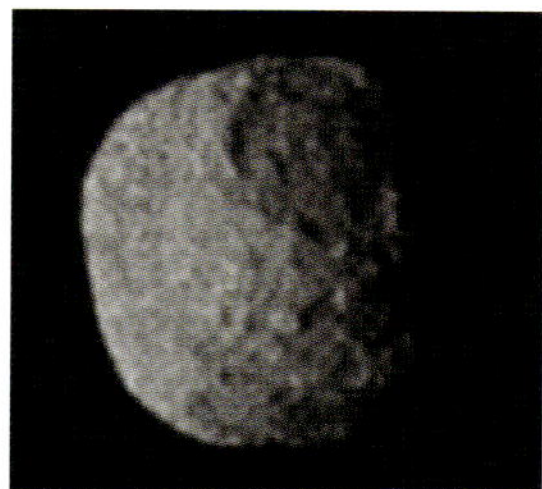
Proteus has a very irregular shape as seen in this photo taken by *Voyager 2*.

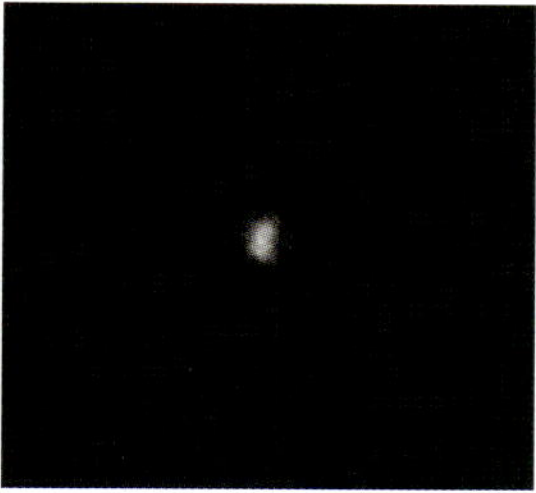
This fuzzy image of Nereid was taken by *Voyager 2*.

and circles Neptune in the same direction that Neptune rotates. This little moon has an irregular shape and is heavily cratered. It is also one of the darkest objects in our solar system and reflects only 6 percent of the sunlight that hits it.

The third largest Neptune moon is **Nereid** (neer' ee id) and was discovered on May 1, 1949, by Gerard Kuiper. We named a group of comets after him—the Kuiper belt. You will learn more about the Kuiper belt in lesson 13. The temperature on Nereid is estimated to be about the same as on Proteus, or 370 °F below zero. This moon has a diameter of 211 miles. Nereid is so far away from Neptune that it takes 365 Earth days to orbit Neptune. That is one year on Earth! Because only one spacecraft has visited Neptune, there is still much more to learn about Nereid!

# Activity 12.2

## Create a Cartoon of Neptune

You can make it a funny cartoon or simply informational. Perhaps your cartoon will include Neptune making a comment about being the last planet. You could have Neptune in a conversation with another planet. Think of something creative and draw your cartoon in your notebooking journal. Then, use word bubbles above or below Neptune to let us know what Neptune is saying.

## What Do You Remember?

What chemical gives Neptune its blue color? Why was Neptune discovered? What made astronomers think there was another planet beyond Uranus? How long does it take Neptune to revolve around the sun? What was the Great Dark Spot? How many moons does Neptune have? What are the names of Neptune's 3 biggest moons? What are geysers? Is water coming from the geysers on Triton? What spacecraft has visited Neptune? What was your favorite part of this lesson?

LESSON 13

# KUIPER BELT AND DWARF PLANETS

## *wisdom from above*

In addition to Mercury, Venus, Mars, Jupiter, and Saturn, the ancient Greeks counted the Moon and sun as planets. Earth was not considered a planet, but rather the central object around which all the other objects in the solar system orbited. However, with the invention of the telescope in the early 1600s, scientists realized that Earth and the other planets rotated around the sun. Because of the telescope, Uranus was added as a planet in 1781, Neptune was discovered in 1846, and Pluto in 1930. As better scientific instruments are developed, our definition and understanding of planets has changed.

*O Lord, how many are Your works!*
*In wisdom You have made them all.*

Psalm 104:24

In the previous chapters, you learned about our solar system's 8 interesting planets, the sun, Earth's Moon, comets, and the asteroid belt. Now it's time to learn about the **Kuiper** (ky' pur) **belt** and **dwarf planets**. Much of what we know about these planets is brand new information. Let's go explore these new and exciting discoveries together!

# God's Creativity

God is the Great Creator, and our incredible solar system is just one example of His creativity. Humans have been studying the solar system for thousands of years. Because it is so large, scientists and astronomers have struggled to see and understand what objects exist beyond Neptune. But now, thanks to unmanned spacecraft, we are able to explore farther reaches of space.

You see, when scientists discovered the Kuiper belt and **Pluto** (plu' toe), we only knew what we could see from telescopes here on Earth. We couldn't send spacecraft to get better pictures and information. Today, scientists have more powerful telescopes and unmanned spacecraft to explore our solar system.

# What is a Planet?

Pluto was once called the ninth planet. Now that we have better ways to study distant objects, we have changed our understanding of what a planet is. Today, we classify Pluto as a dwarf planet. Have you ever heard of a dwarf planet? We're going to have some fun exploring them. Just keep in mind that as better information becomes available, things could change. We are still trying to understand what God already knows!

# The Kuiper Belt

The Kuiper belt is named after the astronomer Gerard Kuiper. In 1951, he predicted the presence of a belt of icy objects beyond Neptune that would explain some mysteries about comets. We say he **predicted** because the Kuiper belt is so far from Earth that Kuiper did not have telescopes powerful enough to see that far. You might think that it is hard to make a prediction, but you are probably good at making them. Think about it. If you are in the house and hear thunder, you will probably predict rain. You base your prediction on what you know, not on seeing the clouds outside. Gerard Kuiper used the information he knew to predict what he could not see in the Kuiper belt.

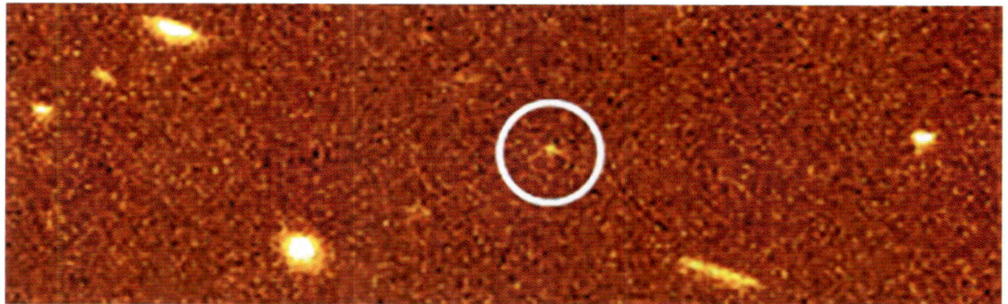

An image of QB1 (indicated by the circle) captured by Jewitt and Luu

In 1992, 2 astronomers, David Jewitt and Jane Luu, found the first Kuiper belt object in real life. They gave it a bit of a funny name: QB1. Scientists believe there are thousands of big objects and millions of small objects in the Kuiper belt.

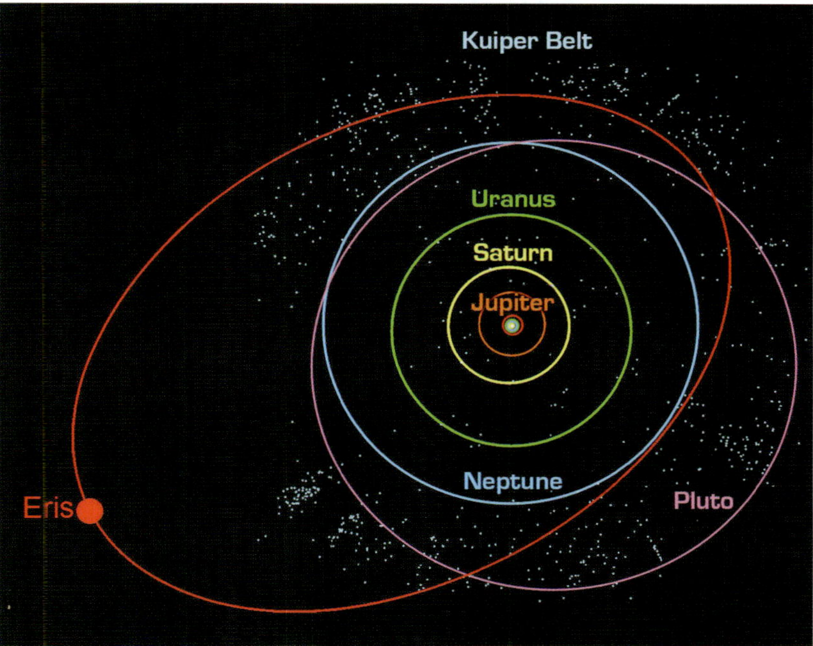


A diagram showing where the Kuiper belt exists in our solar system.

Think of the Kuiper belt as a donut-shaped ring that starts just beyond the orbit of Neptune. The belt is very similar to the asteroid belt except that the objects in the Kuiper belt are usually larger and mostly made up of ice and not rocks. In 2009, NASA's Hubble Space Telescope discovered the smallest object ever seen in the Kuiper belt. It was 3,200 feet across—about the size of a comet. Just as astronomer Gerard Kuiper believed, scientists think that the Kuiper belt is a source of our solar system's short-period comets. Do you remember what a short-period comet is? In addition to short-period comets, the Kuiper belt also includes many dwarf planets.

On January 19, 2006, NASA launched the first spacecraft mission to explore the Kuiper belt. The *New Horizons* mission arrived at Pluto in July of 2015. As it passed by Pluto, it took detailed pictures. It also measured Pluto and studied the strength of solar wind

An artist's concept of *New Horizons* as it passes by Pluto.

this far from the sun. After passing by Pluto, *New Horizons* continued farther into the Kuiper belt where it will continue to study other objects.

Be sure to check **www.apologia.com/bookextras** to keep up to date on all of *New Horizon*'s discoveries.

**Who is the Kuiper belt named after? What did he predict? How do scientists describe the shape of the Kuiper belt? What kinds of objects do scientists believe are in the Kuiper belt?**

## Math It

The Kuiper belt is very far from the sun. Do you remember that the Earth is 93,000,000 miles from the sun? Well, the Kuiper belt starts at a distance 30 times further away from the sun than Earth and ends at a distance 50 times further from the sun than Earth. This distance makes it very difficult for scientists to gather information on the details of this belt. To understand how far away from the sun the Kuiper belt is, calculate the distance in miles to the beginning of the Kuiper belt if it starts 30 times farther than 93,000,000 miles from the sun.

## Dwarf Planets

In 2006, the International Astronomical Union voted on the scientific definition of the terms **planet** and **dwarf planet**. Several things changed when scientists created these official definitions. It was at this moment that Pluto's definition in our solar system changed from planet to dwarf planet. It was also at this moment when an object originally considered an asteroid in the asteroid belt became a dwarf planet named Ceres. We will learn more about these 2 dwarf planets in a little bit. Let's begin by learning the difference between a planet and a dwarf planet.

The International Astronomical Union defines a **dwarf planet** as a body which:

1. Orbits the sun.
2. Has enough mass to assume a nearly round shape.
3. Is not a moon.
4. Has not cleared the neighborhood around its orbit.

An artist's concept of objects that have "not cleared the neighborhood around their orbit" as they revolve about the sun.

What does all this mean? First, the object must orbit the sun. Then scientists look to see if the object is round. When an object is big enough, it has the gravity necessary to pull it into a round shape. Third, dwarf planets do not revolve around another object (they are not a moon). Finally, to "clear the neighborhood around its orbit" means that there are no similar bodies near the planet. This is the primary difference between a planet and a dwarf planet. The strong gravity of a planet pulls tiny space objects into its atmosphere and clears its orbit of debris. Dwarf planets do not have enough gravity to clear their surrounding area of space clutter.

Do you remember from the previous lessons in this book each of the 8 planets we studied have orbits that are unique? That means, as Earth circles the sun it is not in danger of hitting Venus or Mercury. Jupiter orbits the sun, without crashing into Saturn. Dwarf planets, however, orbit the sun with many objects crossing their orbital path.

The first 5 recognized dwarf planets are **Ceres** (seer' eez), **Pluto**, **Eris** (ee' ris), **Makemake** (mah' key mah' key), and **Haumea** (how' may a). Scientists believe that there are many more dwarf planets awaiting discovery. Now that you have a better understanding of what a dwarf planet is, let's explore these dwarf planets in detail.

**Explain to someone what you have learned about dwarf planets before moving on.**

# Activity 13.1

## What Do Dwarf Planets Look Like?

Dwarf planets are very cold places and of ice and rock. The amount of ice and rock depends on the dwarf planet's distance from the sun. A dwarf planet closer to the sun will have more rock showing on the surface while a dwarf planet farther from the sun will have more ice covering the surface. Let us make 2 different types of dwarf planet models, describe how they look, and determine which one would be closer to the sun and which farther.

### You will need:

- Adult supervision
- 2 small balloons
- 10 inches of string
- 2 small rocks (1 inch in diameter)
- 3 cups water
- 1/8 cup dirt
- Eye dropper
- Pie plate
- Magnifying glass (optional)

### You will do:

1. Mix the dirt with 1 cup of water.
2. In one balloon, carefully place one of the rocks and pour in one eyedropper of the water mixed with dirt. Tie the balloon, shake, and place in the freezer overnight.
3. Tie the string around the other rock.
4. In the other balloon, carefully place the rock with the string tied around it into the balloon. Slowly pour into the balloon the 2 cups of water. Adjust the rock with the string so that it is in the middle of the balloon with the water surrounding the rock. Tie the balloon and hang in the freezer overnight. The balloon should be round in shape.

**(Continued on next page.)**

5. After the water freezes in the balloons, take them out of the freezer and remove the balloons. These are the dwarf planet models. Place each dwarf planet model on a pie plate.

## Discussion

Can you see the rock in both dwarf planet models? In which model is the rock more visible? What are some of the unique ice structures you are observing in each model? Is the ice clear, or dark? Why is the ice clear or dark? Draw a picture of both dwarf planet models. Label which would be closer to the sun and which one would be farther from the sun. Explain why you labeled the models as you did.

Share your ice drawings and findings with your family. Explain that dwarf planets consist of rock and ice. Explain why the model with more of the rock showing and a darker ice would represent a dwarf planet that is closer to the sun. Explain why the model with less of the rock showing and clearer ice would represent a dwarf planet that is farther from the sun.

# Ceres

When discovered in 1801, everyone thought Ceres was a planet between Mars and Jupiter. Later, scientists decided it was an asteroid. Ceres remained classified as an asteroid until 2006 when scientists realized it fit the definition of a dwarf planet. Ceres is now defined as a dwarf planet in the asteroid belt. Remember that even though Ceres is in the asteroid belt, all of the other known dwarf planets are in the Kuiper belt. This makes Ceres a unique dwarf planet.

A photograph of Ceres taken by the *Dawn* spacecraft in February 2015.

Ceres is round and smaller than Earth's Moon. Its diameter is only 600 miles across. Scientists think it has a dusty, icy outer surface and a rocky inner core. Using the Herschel Space Observatory, scientists have found evidence of a water vapor atmosphere that comes from the sun warming up the icy surface.

A day on Ceres lasts only 9 Earth hours. That's a pretty short day. Its orbit is quite long, however. It takes Ceres 4.6 Earth years to circle the sun.

The force of gravity on Ceres is much less than Earth. If you weighed 100 pounds on Earth, you would only weigh 2.8 pounds on Ceres. Although you would weigh less on Ceres and could jump higher, it would be very cold. That's because the average temperatures are -158 °F. Would you like to visit that cold little dwarf planet? I wouldn't. Brrr.

An artist's concept of NASA's *Dawn* spacecraft heading toward the dwarf planet Ceres.

Scientists are still learning about Ceres. All of the current knowledge we have about this dwarf planet comes from telescopes. In 2007, NASA launched an unmanned spacecraft, *Dawn*, to study Ceres. *Dawn* arrived at Ceres in February 2015, and sent back pictures and lots of information. It is exciting to see better pictures of this dwarf planet's surface and to learn more about the icy temperatures. There is so much more to learn about Ceres!

**Describe some unique things about Ceres.**

## Pluto

Pluto was discovered in 1930 by an astronomer named Clyde Tombaugh. Until 2006, Pluto was the ninth planet in our solar system. We realize now that it's actually an object inside the Kuiper belt! So, in 2006, scientists reclassified Pluto a dwarf planet.

A picture of Pluto taken by *New Horizons*.

After studying this dwarf planet with telescopes, scientists believe that Pluto is just a little smaller than Earth's Moon. It's about 1,430 miles around with a rocky core covered by an icy surface. We believe that the atmosphere on Pluto is made of methane and nitrogen. However, when Pluto travels furthest away from the sun during its orbit, it gets so cold that scientists think the atmosphere completely freezes. This means that the atmospheric gas turns into a solid and falls to the surface. Can you imagine that?

A day on Pluto lasts 153 Earth hours. That's a long day! Did you know that Pluto rotates in the opposite direction of Earth? Do you remember which other planets do that? It takes 248 Earth years for Pluto to orbit the sun. That's because it's so far away. Since it takes so long to orbit the sun, how would we celebrate birthdays on Pluto if we lived there? We might just have one birthday party when we're born. How would you like that?

The force of gravity on Pluto is much less than on Earth. If you weighed 100 pounds on Earth, you would only weigh 6.7 pounds on Pluto. Pluto is even colder than Ceres, with an average temperature of -400 °F. The coldest temperatures on Earth only get to -136 °F. No wonder the atmosphere freezes.

Size comparison of Earth, Earth's Moon and Pluto.

## Strange Orbit

The way Pluto orbits the sun is not even. Sometimes it orbits closer to the sun; and sometimes closer to Neptune. The last time that happened was from 1979 to 1999. Another strange thing about Pluto's orbit is that it doesn't travel in the same plane as the other planets in our solar system. Pluto's orbit tilts. This makes Pluto's revolution around the sun unique from all of the other planets in our solar system. Some scientists believe that this strange orbit resulted from a comet crashing into Pluto years ago. However, no one knows for sure.

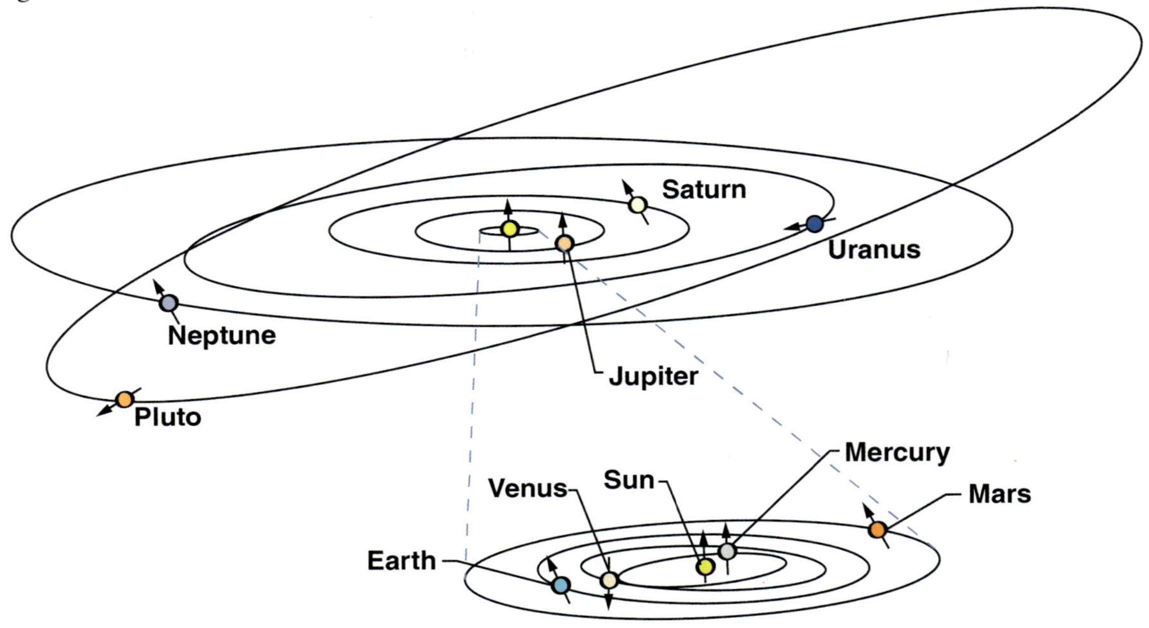


This diagram shows Pluto's orbit around the sun in relationship to the other planets. Do you notice how Pluto's orbit tilts?

## Pluto's Moons

Even though Pluto is smaller than the other planets in our solar system, scientists have discovered that it has 5 moons! Its largest moon is **Charon** (shair' uhn) and is about half the size of Pluto. The other 4 moons are **Hydra** (hi' druh), **Styx** (stiks), **Nix** (niks), and **Kerberos** (kur' ber uhs).

It took scientists almost 50 years after the discovery of Pluto to find Charon since it is so small and so far away from Earth. Charon stays above the same spot on Pluto's surface all of the time. Although Charon is half the size of Pluto, it is quite large for a moon. Because of this, some scientists describe Pluto and Charon as a **double dwarf planet system**.

In 2005, scientists discovered Pluto's moons Nix and Hydra using the Hubble Space Telescope. They believe that these 2 little moons, each about 100 miles wide, may have formed at the same time as Charon. In 2011, again using the Hubble Space Telescope, scientists discovered the fourth moon, Kerberos. Scientists discovered the fifth moon, Styx, while searching for objects in the path of the *New Horizons* spacecraft. Little else is known about these 4 small moons. Do you think they'll find any more moons around Pluto?

This illustration depicts Pluto and 5 of its moons looking away from the sun.

**Using your own words, explain what you learned about Pluto and its moons.**

# Eris

In 2005, astronomer Michael Brown, and his colleagues Chad Trujillo and David Robinowitz, discovered Eris. Scientists studying Eris have calculated its diameter to be 1,445 miles, slightly larger than Pluto. This larger diameter is what triggered the debate in the scientific community that eventually lead the International Astronomical Union to define planet and dwarf planet. Therefore, it was the discovery of Eris that resulted in Pluto being renamed a dwarf planet!

Eris is so far away from Earth and so small, it is not easily visible—even with telescopes! However, data shows that Eris has a presence of methane ice, which means it may look similar to Pluto. We still do not know if Eris has an atmosphere.

A day on Eris lasts about 24 Earth hours and rotation is the same as on Earth. This means that the sun rises in the east and sets in the west. Eris orbits the sun with a maximum distance of 8,928,000,000 miles and takes 557 Earth years to orbit the sun!

A picture of Eris and its moon, Dysnomia.

The force of gravity on Eris is much less than on Earth. If you weighed 100 pounds on Earth, scientists estimate that you would only weigh 8 pounds on Eris! It is very cold with average surface temperatures estimated at -400 °F. Remember that the coldest temperatures on Earth only get to -136 °F. Like the other dwarf planets, Eris is a cold place.

Eris orbits the sun such that during its revolution, it extends far beyond the Kuiper belt. Unlike Pluto, whose orbit sometimes brings it closer to the sun than Neptune, Eris's orbit always remains farther from the sun than Neptune. Other than the fact that Eris has a small moon named **Dysnomia** (dis' no mia), very little else is known about Eris at this time.

# Makemake

Makemake was discovered on March 31, 2005, by a team led by astronomer Michael Brown, the same person who discovered Eris. Scientists studying Makemake calculate its diameter to be about 900 miles, a little smaller than Pluto.

Makemake as photographed by the Hubble Space Telescope.

Makemake is not clearly visible from Earth's telescopes. However, data shows that Makemake has a presence of methane and nitrogen that tells us it may look similar to Pluto. So far, no atmosphere has been detected on Makemake.

Scientists believe that a day on Makemake lasts about 22.5 Earth hours and it rotates the same direction as Earth with the sun rising in the east and setting in the west. Makemake orbits the sun at a maximum distance of 4,863,900,000 miles and takes 310 Earth years to orbit the sun.

Scientists believe that like other dwarf planets, the force of gravity on Makemake is much less than on Earth. If you weighed 100 pounds on Earth, scientists estimate that you would only weigh 5.1 pounds on Makemake. It is very cold with average surface temperatures estimated at -406 °F. Remember, the coldest temperatures on Earth only get to -136 °F. Makemake is another really cold place. Makemake is also visually the second-brightest Kuiper belt object after Pluto. Scientists have not yet found a moon orbiting Makemake. However, there is still much to learn about this incredible dwarf planet!

## Haumea

Although discovered in March 2003, the official announcement of Haumea's discovery didn't occur until 2005, the same year that its two moons were discovered. Scientists believe Haumea is about 807 miles across. That's about the distance from the top of California to the bottom of California. That's pretty small. So that's why it can't be clearly seen with telescopes. However, scientists believe Haumea is a rocky dwarf planet covered with a thin layer of ice and has no atmosphere.

Haumea rotates in the same direction as we do here on Earth, but a day lasts about 3.9 hours. This is the fastest of any planet in the solar system and is most likely why the shape of Haumea is slightly oval like an egg. Haumea orbits the sun at an average distance of 3,990,000,000 miles and takes 283 Earth years to orbit the sun.

Because it's so small it doesn't have the same force of gravity as Earth does. So, if you weighed 100 pounds on Earth, you would only weigh 4.5 pounds on Haumea. Haumea is very cold, with average surface temperatures estimated to be -307 °F. Like all other dwarf planets, Haumea wouldn't be fun to visit.

## Haumea's Moons

Two small moons revolve around Haumea. They are named **Hi'iaka** (hi uhi a' ka) and **Namaka** (na ma' ka). Hi'iaka, was discovered on January 26, 2005. It is the larger and brighter of the 2 moons with a diameter of 193 miles and an orbit of 49 Earth days. Scientists believe that ice covers much of the moon's surface. Discovered later that same year, on June 30, 2005, Namaka is a smaller moon with a diameter of 106 miles. Namaka orbits Haumea in 18 Earth days.

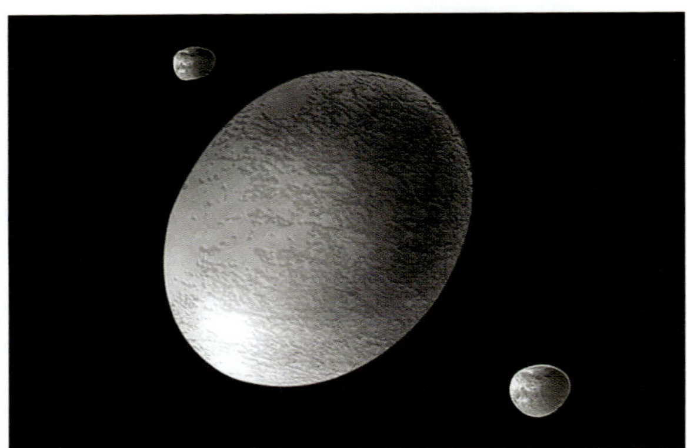

An artist's concept of what Haumea and its moons may look like.

**Take a moment to tell someone at least one interesting fact about each dwarf planet.**

## Dwarf Planets in Review

There are currently 5 known dwarf planets in our solar system. Ceres is the only known dwarf planet in the asteroid belt. The other 4 dwarf planets, Pluto, Eris, Makemake, and Haumea are located in the Kuiper belt. The picture to the right shows the relative location of these dwarf planets in our solar system.

All of the dwarf planets are smaller than Earth's Moon. The smallest dwarf planet is Ceres and the largest is Eris. A comparison of their sizes is shown on the next page.

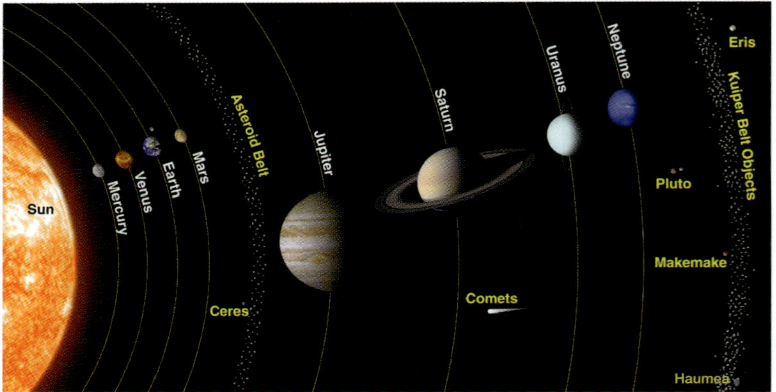


Our solar system.

Let us take a moment to review what scientists currently know about dwarf planets.

1. If the sun were the height of a standard front door, Earth would be the size of a nickel, and dwarf planets Eris and Pluto would be the size of the head of a pin.
2. Dwarf planets orbit the sun. Most are located in the Kuiper belt. Ceres is located in the asteroid belt.
3. Pluto is the most famous dwarf planet.
4. Dwarf planets are solid objects composed of rock and/or ice. The amount of rock or ice depends on their location in the solar system.
5. Many of the dwarf planets have moons.
6. There are no known dwarf planets with rings like the ones around Saturn.
7. Scientists believe that 2 dwarf planets (Pluto and Ceres) have atmospheres.
8. The first spacecraft mission to a dwarf planet is *Dawn*.
9. The first spacecraft to visit Pluto and the Kuiper belt is *New Horizons*.
10. Dwarf planets are cold places.

You now have a better understanding of what a dwarf planet is. And yet, there is still much to learn about these wonderful celestial objects. Maybe you will be part of a scientific team that learns more about the dwarf planets we know, or even discovers more dwarf planets within our solar system.

# Activity 13.2

## The Earth vs. Dwarf Planets

In this activity, you will make a chart that shows the similarities and compares the differences between Earth and the dwarf planets.

| Similarities Between Earth and Dwarf Planets | Differences Between Earth and Dwarf Planets |
| --- | --- |
| | |

After you fill out the chart, answer these questions:

1. How are the Earth and the dwarf planets similar?
2. How are the Earth and the dwarf planets different?
3. Are there similar characteristics between the 5 dwarf planets? If so, what are they?
4. Do you believe people could live on a dwarf planet? Explain your answer.
5. Which planet or dwarf planet has the most moons?

## What do you remember?

What is the Kuiper belt? What are the differences between a planet and a dwarf planet? Why do we use spacecraft to study the dwarf planets? What was your favorite part of this lesson?

LESSON 14

# STARS, GALAXIES, AND SPACE TRAVEL

*wisdom from above*

"He counts the number of the stars; He gives names to all of them."

Psalm 147:4

The brightest star in this photo is called Deneb.

## Star Light, Star Bright

We are now going to study what is beyond our solar system. The night sky is sparkling with many tiny stars, some dim and some bright. Some of the stars we see in the sky are actually planets, but without a telescope, they look like stars. These stars surround Earth all of the time, day and night, but our sun's light is so bright that it drowns out the light from the stars during the daytime. We can only see the stars when Earth is facing away from the sun. The stars are farther away than the planets. They are outside of our solar system.

What is a star? A star is a large celestial body that produces so much energy that we can see its light, even though it is far, far away. Some stars are farther away than others. Our sun, remember, is a star. There are many billions of stars in the universe. New ones are discovered each day. Some have been named by man; billions more have not. Nevertheless, God said He calls each star by name.

On a cloudless night, when the Moon is new (completely dark), you can see about 3,000 stars in the sky. With a fairly inexpensive telescope, you can see about 100,000. As we gaze at the stars, they appear to twinkle. That's because the particles in the atmosphere surrounding Earth are busily moving around. Every few seconds, these particles pass between a star and our eye. When this happens, the star dims for a quick second as the particle passes in front of it. The star then brightens again when the particle moves away. This makes it look like the star is twinkling. As man-made satellites orbit our Earth (like the Hubble Space Telescope), they take pictures of the stars. Since they are outside of our atmosphere, they get very clear pictures.

Different stars are visible in the night sky as we revolve around the sun and move through the 4 seasons. This means we get a different view of the vast universe every few months! One of the most important stars, besides our sun of course, is the **North Star**, which is also called **Polaris** (poh lair' us). The North Pole is always pointing toward Polaris, so even as all the other stars we see change, those who live in the Northern

Hemisphere can always see the North Star. Also, since we know that the North Pole points toward Polaris, if you can see Polaris, you know which way is north. Because of this, Polaris was an important navigational tool during early exploration.

Polaris is part of a group of stars called the **Little Dipper**. Although many people call the Little Dipper a **constellation** (kahn stuh lay' shun), it is not; it is an **asterism** (as' tuh riz uhm)—a set of stars forming a shape that is easy to recognize. As you see in the photo, Polaris and its surrounding stars form what looks to be a spoon with a curved handle. That's the Little Dipper. It is called the Little Dipper because another set of nearby stars form a bigger spoon shape. Not surprisingly, we call it the **Big Dipper**. The two stars which make up the edge of the Big Dipper's ladle form a line that points to Polaris. When you learn about constellations later on in this lesson, you will learn that the Big Dipper is a part of the constellation known as Ursa (ur' suh) Major, and the Little Dipper is a part of the constellation known as Ursa Minor.

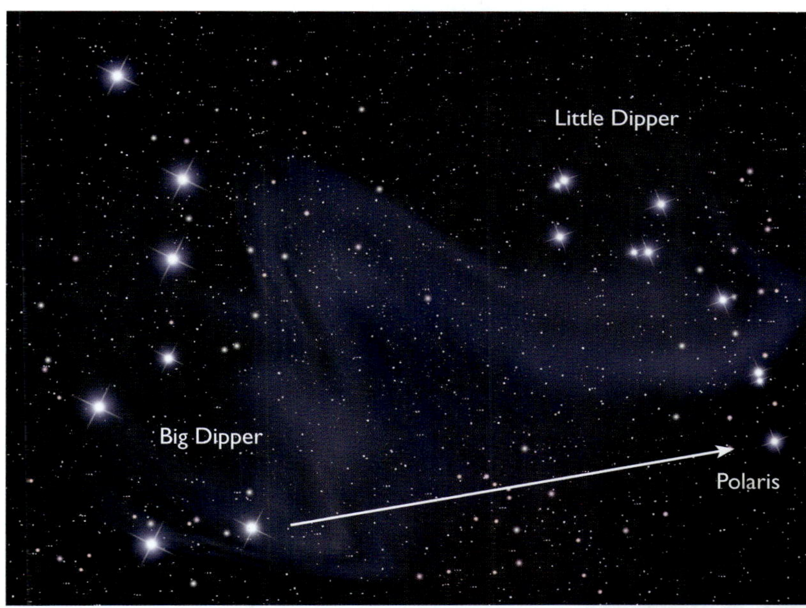

A photo of the Big Dipper and the Little Dipper. The ladle of the Big Dipper points to the North Star, which is on the handle of the Little Dipper. A photographic technique was used to make the stars stand out from the background. In real life, the asterisms are not this bright.

As the stars journey through the sky, all the stars appear to rotate around Polaris. To understand this concept we will do an activity.

# Activity 14.1
## Understanding the Night Sky

### You will need:
- Adult supervision
- Dark colored umbrella
- White chalk

### You will do:
1. Open your umbrella and look inside. Do you see that there are lines that divide the umbrella into sections? Those lines are like the segments that astronomers use to divide up the night sky. Every segment represents a different month.
2. The very tiptop where the pole comes through the umbrella is where Polaris is located. Take your chalk and make a mark around the center to make Polaris.
3. Next, use your chalk to make star patterns around the inside of the umbrella.

**(Continued on next page.)**

4. You can spin the umbrella around and watch all the stars move around Polaris, the center.
5. Focus your eyes on one segment of the umbrella. Within that segment are the stars you are able to see from Earth that month.
6. When you turn your umbrella, different stars come into view. The North Star, Polaris, remains in the same spot!
7. If you look up at Polaris as you turn your umbrella, it looks as if all the stars are turning around Polaris. That is what it is like to view the stars from the Northern Hemisphere of Earth.

The brightest star in the sky is not the North Star. It is **Sirius** (seer' ee us). Sirius is mentioned and described in ancient writings from 1,500 years ago. Sirius, like many stars in the sky, is actually a **double star system**, which is also called a **binary star system**. That means that there is another star in the solar system of Sirius that orbits around Sirius. The star that orbits around Sirius is dim and small. It is a **white dwarf star**.

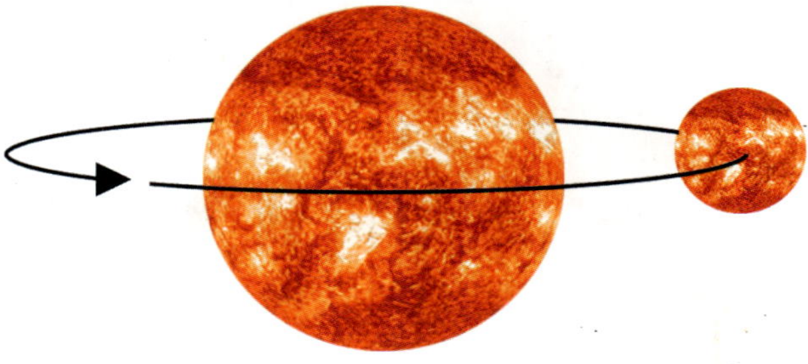

A double star system

White dwarf stars are actually stars that are in the process of dying. Stars die? Yes, they do. Stars have a limited lifetime because the fuel that they use for thermonuclear (thur' moh new' klee ur) fusion eventually runs out. Do you remember what thermonuclear fusion is? You learned about it in lesson 2. Thermonuclear fusion takes fuel. When that fuel runs out, the star begins to die, falling in on itself and becoming smaller. For certain stars, this can actually form a **black hole**. The Bible says that this world is passing away (Matt 24:35). Star deaths point to a decaying universe, just like the Bible says.

**Take a moment to tell someone what you are learning about so far in this lesson.**

An artist's idea of gases spinning around a black hole. You cannot see the black hole; it is at the very center of the swirl. However, you can see all of the activity going on around it, and from the activity, you can conclude that the center is, indeed, a black hole.

## Black Holes

When a star runs out of fuel, it begins to collapse inward. This makes it get smaller and smaller. What makes a dying star collapse like that? The star's own gravity pulls it inward. When this happens, the star begins to shrink. This heats up the core of the star. If the star has a lot of mass, the heat generated by the collapse is truly awesome, and the star explodes! This is called a **supernova** (soo' per noh' vuh). When the star explodes, part of the star gets shot into space, but part of the core stays behind. At that point, the core is very small and has a *lot* of mass. Do you remember what mass does? It causes gravitational pull. The core that is left behind is so small and has so much mass that its gravity pulls everything, even light, into it. At that point, the object is dark because light cannot escape it, and we call it a **black hole**. Strange, huh?

We can't see black holes because they are black, and to see something with a telescope, light must shine from it. However, the strong gravity of a black hole affects the things around it. When stars or gases behave in peculiar ways, that can indicate the presence of a black hole nearby. For example, we can see gases in space that orbit very quickly in tight circles. The speed at which they are moving and the size of their orbit suggests that they are orbiting a black hole.

Don't worry about Earth being sucked into a black hole. Because the universe is expanding (getting bigger and bigger), we are moving farther away from any nearby black holes. Think about what happens to a balloon as you blow it up—it gets bigger and bigger. That is what is happening to the universe. All the stars are moving farther and farther apart from one another. We are getting farther away from every star, black hole, and solar system that surrounds us.

# Activity 14.2
## Understanding the Expanding Universe

### You Will Need:

- Adult supervision
- Balloon
- Pen or marker
- Measuring device, such as a ruler or tape measure

(Continued on next page.)

### You Will Do:

1. Draw little stars all over your balloon with the pen or marker.
2 Measure how far apart the stars are from one another.
3. Blow up the balloon and tie it.
4. Measure the distance between the stars again.

### Discussion

When the balloon expanded, the stars became farther apart, didn't they? As the universe expands, we are getting farther away from other stars just as the stars on your expanding balloon moved away from each other.

**Take a moment to explain to someone what a black hole is and how our universe is expanding.**

A photograph of the Crab Nebula, which contains the remains of the first recorded supernova.

## Supernovas

When I told you how a black hole forms, I mentioned that the collapse of some stars can lead to giant explosions, called supernovas. Believe it or not, we have observed this phenomenon! In February 1987, 3 separate astronomers were surprised to see a bright new star. When they compared it to previous images of the same part of the sky, they saw that no bright star had been there before. The bright star remained there for several months and slowly dimmed away. It was quickly determined that the star was a supernova, and it was named Supernova 1987A. This was not the first appearance of a supernova, however. Supernovas have been seen throughout history. One of the earliest recorded ones was seen by Chinese astronomers in 1054. That supernova was so bright that it was visible in the daytime!

As beautiful as a supernova is, what they leave behind is even more wonderful. The gases and dust thrown from the exploding star form a cloud of dust and debris called a **nebula** (neb' yuh luh). Nebulae (plural of nebula) have all sorts of interesting shapes and colors, depending on the type of star that went supernova. An example of a nebula is shown above. The Chinese astronomers who observed the supernova in 1054 recorded where it appeared in the sky. When later astronomers pointed a telescope in that direction, they found the **Crab Nebula**. Thanks to the detailed records of Chinese astronomers from nearly 1,000 years ago, we can see the nebula that is the remnant of the supernova that they saw.

**Use your own words to describe a supernova to someone.**

# Variable Stars

Unlike many stars, our sun puts out nearly the same amount of heat, light, and energy every day. When measured over a long period of time, scientists have determined that the sun's heat, light, and energy is increasing (as you learned in lesson 2). But on a day-to-day basis, the energy that comes from the sun is consistent. We never wake up in the morning wondering if the sun will dry up the oceans before the day is over. We don't go to bed worrying that the sun might become so dim that the whole world will freeze while we sleep. Believe it or not, this is exactly what some other stars do. Sometimes they burn very hot and bright; at other times, they dim, putting out less heat and energy. These stars are called **variable stars** because the energy in them varies.

Do you remember what a binary (or double) star system is? It is a system in which one star orbits another star. Well, nearly half of all of the stars you see in the sky are double stars! Given all the double stars and all the variable stars in the sky, you can see how special our own star is! In fact, most stars in the universe have many features that would make it impossible for us to live on a planet that orbited them. We could not survive if our sun was not just the way God made it. Aren't you glad that God loves us enough to create the perfect star to orbit?

An artist's concept of a binary star system.

## Categorizing Stars: *Hot or Cold*

Scientists can tell how hot a star is by its color. Have you ever watched a fire burn in a fireplace or at a campsite? The coals at the bottom of the fire glow red when first lit, but as the coals get hotter, they burn white. The color of a star also tells you how hot it is. Blue stars are the hottest stars, while red stars are the coolest. Yellow stars, like our sun, fall in the middle of the temperature range.

Sometimes, when organizing a group of people, a leader will categorize them based on the first letter of their last name. What is your last name? If your last name is Smith, you would be in the S category. This would be arranging the group by letter. That's what scientists do with stars—once the heat of a star is determined, the star is put in a letter category with other stars of similar temperature. There are seven categories for stars: O, B, A, F, G, K, and M. The letter *O* stands for the hottest stars (the blue ones), and *M* stands for the coolest stars (the red ones). Our sun is a *G*, a medium-temperature star.

The way astronomers remember the order of these letters is by creating a mnemonic. Do you remember the mnemonic to learn the order of the planets? Here is another silly sentence: "Of Berkeley Astronomers, Few Give Kind Marks." Berkeley is a very tough college, and *marks* is another way of saying *grades*. This sentence is easy to remember for anyone who studied astronomy at Berkeley! Do you see the first letters of the words in that sentence? They are O, B, A, F, G, K, and M. That's the order of star categories from hottest to coolest.

# Activity 14.3

## Make a Mnemonic

Make a mnemonic to help you remember the star categories based on temperature.

| O | B | A | F | G | K | M |
|---|---|---|---|---|---|---|
| | | | | | | |

Place your mnemonic in your notebooking journal.

**Take a moment to tell someone about variable stars and how stars are categorized.**

## Categorizing Stars: *Bright or Dim*

What grade are you in? If you are in first grade, you are classified under the number 1. If you are in fourth grade, you are classified under the number 4. If your last name is Smith and you are in the fourth grade, you could be put in the category S-4, meaning that your last name starts with S and that you are in the fourth grade. Stars are also assigned a number as well as a letter. The number tells you how bright the star is. Oddly enough, a smaller number means a brighter star. A 1 star is much brighter than a 9 star. When you put the letter and the number together, you learn a lot about a star. Our sun, for example, is a G-2—the G means it is a medium-temperature star, and the 2 means that it is dimmer than Sirius, which is an A-1 star. Do you remember what Sirius is? It is the brightest star we see in the night sky. Wait a minute. How can the sun be dimmer than Sirius? The sun is so bright that it drowns out all of the stars, including Sirius. Well, when astronomers talk about how bright or dim a star is, they do not mean how bright or dim it looks from Earth. They mean how bright or dim the star is when you are close to it. The sun looks very bright to us because we are so close to it. If the sun and Sirius were both the same distance from Earth, Sirius would look *much* brighter.

## Categorizing Stars: *Big or Small*

There is an incredibly large star not too far from Earth called **Betelgeuse** (beet' uhl jooz). It is so big that if it were sitting in the middle of our solar system, it would take up all of the room from the sun to Jupiter! That's many, many times bigger than our sun. In fact, it is so big that astronomers call it a **supergiant** star. You can find Betelgeuse at the top of the constellation **Orion** (oh rye' uhn). Orion is called *the hunter* because if you connect the stars and use your imagination, it kind of looks like a man holding a shield (or maybe a bow).

Stars not only get labeled according to heat and brightness, they are also graded on the total amount of energy they emit. In general, bigger stars make more energy than smaller stars, so these designations also tell you how big the star is. The grades are given with Roman numerals, and Betelgeuse gets the highest grade.

- I: supergiant stars (Betelgeuse is a I.)
- II: bright giant stars
- III: giant stars
- IV: subgiant stars
- V: main sequence stars (Our sun is a V.)
- VI: sub-dwarf stars
- VII: white dwarf stars (The star that orbits Sirius is a VII.)

Our sun is a G-2 V star. Do you remember what that means? Now, as you study the stars you will understand the importance of their letters and numbers.

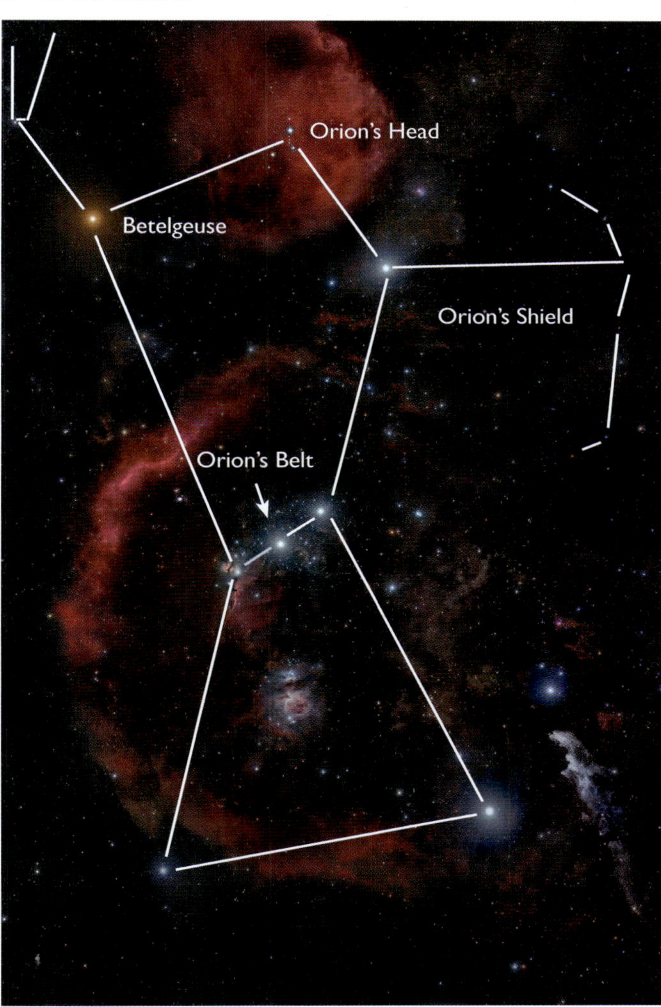

This is a photo of the constellation Orion. When looking at the sky, the shield and head are very hard to see. However, the body, belt, and feet are easy to find. Without the head and shield, it looks like a big K in the sky.

**Put into words all that you remember so far. Don't forget about the North Star, Polaris, and the brightest star, Sirius. Also explain what black holes and supernovas are.**

# Activity 14.4

## Make an Astrometer

In this activity, you will build an **astrometer** (as troh' meh tur) that will measure the brightness of the stars according to the numbers given stars by astronomers. Your astrometer will measure grades 1-4. Please note that because we are on Earth, our astrometer will measure the *apparent* brightness of the stars (how bright they are from Earth), which is different than the *actual* brightness of the stars. If you ever study astronomy in college, you will learn how to take the brightness you see here on Earth and turn it into the actual brightness of a star.

### You will need:

- Adult supervision
- Cardboard
- Clear plastic wrap
- Tape
- Scissors
- Marker

### You will do:

1. Cut 4 rectangular slots in the cardboard.
2. Tape a long piece of plastic wrap over all 4 rectangles.
3. Tape another piece of plastic wrap over the 3 top rectangles.
4. Tape another piece of plastic wrap over the 2 top rectangles.
5. Tape another piece of plastic wrap over the top rectangle. Now you have 4 rectangles covered in plastic wrap.
6. Write a **1** next to the rectangle with four layers of wrap. Write a **2** next to the one with three layers, a **3** next to the one with two layers, and a **4** next to the one with only one layer of wrap. You have just made an astrometer. You can use it tonight to measure the brightness of stars.
7. Take your astrometer outside on a clear, dark night, and begin by looking through square number four. You will see many stars through it.
8. Look through square number three. You will see fewer stars this time.
9. Look through number two, and move to number one. The rectangle with the most plastic wrap only shows the brightest stars in the sky.

### Discussion

If you want to determine the brightness of a single star, look at it through each rectangle, starting at rectangle 4 and progressing through the rectangles in decreasing order. The number of the last rectangle through which you can still see the star gives you the brightness of the star.

# Light Years

Did you know that light travels very quickly? Light travels at 186,000 miles per second. Say that out loud. Light travels at one hundred eighty-six thousand miles per second. That means that when you turn on the light, it traveled that fast to light up your room. When you see the light from the sun, the light you are seeing has traveled 93 million miles, going 186,000 miles per second to get here. That means it takes the light from the sun about 8 minutes before it finally gets to your eye. So, when you see the sun each day, you are really seeing what the sun looked like almost 8 minutes ago! You can never see what the sun looks like right this minute unless you get very close to the sun in a spacecraft. If you were on Pluto, the light you would see from the sun would be about 5.5 hours old by the time you saw it!

The stars in the sky are very far away from us. The closest star to us is **Proxima** (prox' ih muh) **Centauri** (sen tor' ee), and it is about 4 light years away from Earth. A **light year** is the distance that light can travel in a year. Since light travels 186,000 miles per second, it can travel a *long* way (about 5,900,000,000,000 miles) in a year. This means if we were going as fast as light (186,000 miles per second), it would take us 4 years to reach the closest star! What's really incredible is that most of the stars in the sky are hundreds, thousands, millions, even billions of light years away from us. Even traveling at the speed of light, we could not visit most of them in our lifetime.

## Galaxies

Stars are found in groups—large groups. These big groups of stars are called galaxies. **Galaxies** are billions of stars clustered together. Galaxies come in four shapes: **spiral**, **elliptical**, **lenticular** (len tik' you lur) and **irregular**.

Earth is in a galaxy called the **Milky Way**. Our galaxy is called the Milky Way because if you go outside on a clear night, you will see a milky white, almost pink, smear across the sky. That smear is one arm of the spiral that makes up our galaxy. Unfortunately, many people cannot see the Milky Way anymore because the light pollution from bright city lights drowns out many stars.

A view of the Milky Way from West Virginia.

A spiral galaxy; called that due to its spiral shape.

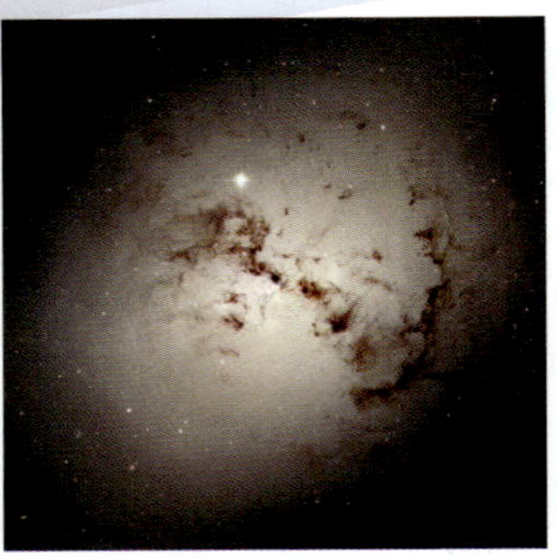
An elliptical galaxy.

A lenticular galaxy. Though similar to a spiral galaxy, it has no arms like the spiral galaxy.

An irregular galaxy. If a galaxy is not a spiral, elliptical, or lenticular galaxy, it is an irregular galaxy.

The Milky Way is a spiral galaxy. Scientists have determined that the only location in which a star could have a solar system capable of supporting life is on the arm of a spiral galaxy—exactly where God placed Earth! In order to support life, Earth needed the perfect size, the perfect rotation, and the perfect environment. Earth also needed to orbit a medium-hot, medium-sized, bright sun. We also had to be the perfect distance from that sun. In addition, we now know that for life to survive on a planet, it must be on the arm of a spiral galaxy. It is amazing that some people think Earth and all it contains is an accident. The chance of an accident like us occurring anywhere in the whole universe is almost impossible!

Galaxies are enormous. So enormous, our mind cannot imagine the size of a single galaxy. Not only are there billions of stars in a galaxy, there are billions of galaxies in the universe. Some are so far away that astronomers can barely see them. Remember that things that are far away seem smaller. Some of the galaxies that astronomers have discovered appear so small that the entire galaxy looks like a speck of glitter on the telescope. It is hard to know exactly what is going on, even in our own galaxy, not to mention a galaxy that is billions of light years away!

**Explain what you know about galaxies in your own words.**

# Constellations

When you look up at the November sky, you will not see the same stars that you saw in May. Every month our view of the stars changes. This is because Earth is revolving around the sun and faces different sections of the universe each month. There are stars in all the universe, and as we orbit the sun, the night sky shows us different views.

Do you remember seeing the pictures of Orion, the Big Dipper, and the Little Dipper? The Big Dipper is part of a constellation Ursa Major (great bear) and the Little Dipper is a part of the constellation Ursa Minor (little bear). The Big Dipper and Little Dipper are much easier to see than the entire constellations, however, so it is best to start out looking for the dippers rather than the bears.

Realize that the constellation pictures you see in this book have been enhanced to make the patterns easier to see. When you look at the night sky, the constellations will not be as easy to see. However, with a little patience and practice, you will be able to find them. You can also get some help by going to **www.apologia.com/bookextras**. There will be links to star maps there that will help you know where to look for each constellation.

The constellation Cassiopeia (ka' see uh pea' uh).
It looks like a lopsided W in the sky.

The constellation Canis Major. It is supposed to look like a dog because "canis major" means great dog. It is easy to find because the brightest star in the sky is in it.

**What are constellations? Can you explain them in your own words?**

# Constellations and Astronomy

Why are constellations still given importance in astronomy today? They help people, both astronomers and stargazers, know where they are looking. In other words, it gives them a map of the sky and helps them to find what they are looking for. If you wanted to find Betelguese, for example, you would first locate the constellation Orion, and then it would be easy to locate Betelgeuse.

Before people had a compass, they knew which direction to travel by land or by sea if they looked up at the constellations. They knew which way was north, south, east, or west just by the stars.

You can purchase star maps or **planispheres** (plan' uh sfears) to help you to find the constellations in the sky. If you do purchase a planisphere, you must get one that is for the region where you live. Typically, planispheres are ordered for a particular **latitude**, which marks how many degrees you are from the equator. If you live in the Southern United States, you will want a planisphere for a latitude of 30° north. That means you are 30 degrees north of the equator. If you live in another part of the United States, find your state on the map below and determine which latitude you are closest to. If you are not in the United States, you can get your latitude from an atlas or online.

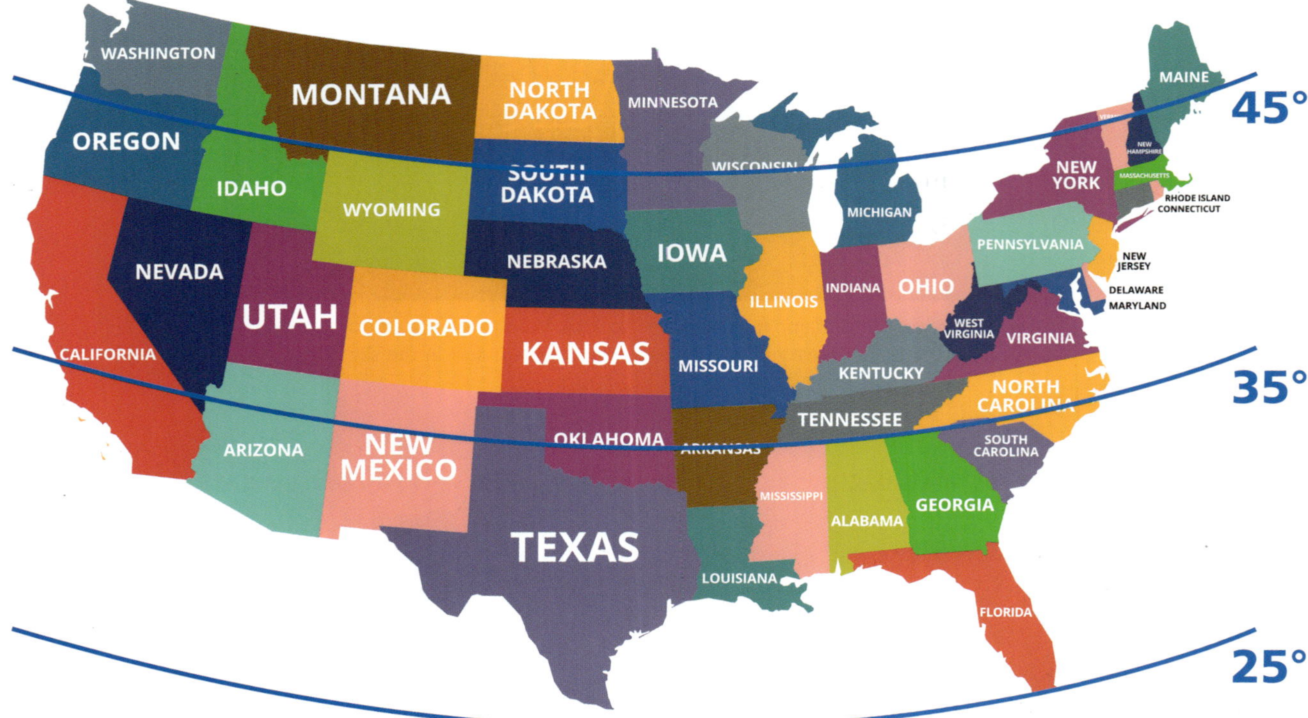

# Activity 14.5
## Locating Constellations

On a nice, clear night, go outside and try to find some of the constellations in the sky. The constellations that you will be able to see will depend on where you live and what time of year it is. To find out what constellations are visible for you, go to **www.apologia.com/bookextras**. There, you will find links to places where you can get star maps for your area and time of year. Those star maps will tell you roughly where to look to find certain constellations.

The best thing to do when you are looking for the constellations is to find someplace very dark. If you live in the city, you might try driving out to a country field or a park to get away from the city lights. Carry a tiny flashlight with you. If you have red plastic, like colored plastic wrap, cover the flashlight with it so that the light will be red. That way, when you use your light, it will not affect your eyes as much.

Before you go out, study your star map for a moment to try to get an idea where you will look for what constellations. Once you get outside, lie down on a blanket or sit in a reclining chair and simply look at the sky for a while. This will get your eyes used to the dark. Then, look for the constellations. If you need to refer to your star map, do so only with the tiny flashlight.

# Let's Go to Space

Since the beginning of time, humans have been fascinated with God's starry hosts, the universe. When humans are interested in something, they usually want to see it up close. For thousands of years, humans have dreamed of visiting the planets. So far, no one has explored another planet—just Earth's Moon. Visiting planets is still a dream. Nevertheless, we have accomplished a lot in our explorations of the universe, and I expect that a lot more will be accomplished in the next generation.

People who study ancient history know that in the 1200s, the Chinese had figured out how to build rockets, but they did not share how to do this with others. Many people tried to make rockets that would blast them up into space. People used dynamite or gunpowder to try to launch rockets into space, but they had no success. In the late 1800s, a Russian astronomer, Konstantin Tsiolkovsky (sil' oh kof' ski), believed that man would need to mix liquid hydrogen and liquid oxygen to make a fuel strong enough to enable a rocket to escape Earth's gravity. No one tried it until many years later, but it did work. Today, we use a very similar mixture to blast rockets into space. Because of this, Tsiolkovsky is called the father of **astronautics** (as truh not' iks). The word *astronautics* means the science of space flight.

Later on, in 1919, a man named Robert Hutchinson Goddard studied how a person might get to the Moon. He told people that a rocket should be used. Everyone laughed at him, made fun of him, and called

Robert Goddard.

him names, like Moon Man. That didn't stop Goddard from working on his rockets. He made them so that they were fast! They could travel with a speed of about 700 miles per hour, and they could go about a mile and a half up into the air before falling back down to the ground. Remember that our atmosphere is about 75 miles high, so Goddard had to keep working to make rockets that would get out of Earth's atmosphere.

About the same time, Germany was in a war and figured out how to make airplanes fly 100 miles per hour and 3 miles up into the air. Mankind was getting higher into the atmosphere. It wouldn't be long before they sent a rocket into space. In fact, many years later, the Russians did just that! Can you find Russia on a map? It's a very large country. Many people back then were frightened of the Russians because their country was not free. People could not live where they chose or work as they chose. They were told what they could and could not believe. The Russian government even tried to forbid the Russian people from believing in God. The government controlled the lives of the people in a way that frightened free nations.

## Sputnik Sensation

The Russians used a lot of ideas that the Germans developed during World War II to make a rocket that would launch into space. The rocket's job was to put the first artificial satellite, called **Sputnik**, into orbit around Earth. This was the first rocket to go to space, and *Sputnik* became the first artificial satellite. The U.S. and other countries were scared that Russia might become very powerful, taking over the whole world with such amazing, advanced technology.

As the United States was working hard to build a rocket of its own, Russia put another rocket into space. *Sputnik 2* was even more amazing because it carried a little dog into space with it. The dog was named **Laika** (like' uh), which is the Russian word for barker. Laika lived in a pressurized cabin—it had an artificial environment with oxygen and heat. Before Laika went to space, many scientists believed that living creatures could not survive without gravity and the protection of our atmosphere. Laika showed the world that living things could survive in space. The Russians were really doing amazing things!

A photograph of *Sputnik 2* prior to be launch.

## Space Race

*Sputnik* created great fear in the United States and other free nations. In response to all the Russians were doing, the United States government poured more funding into the space program to try compete with the Russians. They called this the **Space Race** between the U.S. and Russia. This is what started all that we know today about space and space travel. Just a few months after *Sputnik*, the United States launched *Juno 1*, a rocket carrying a satellite called *Explorer 1*. Finally, a nation besides Russia had a satellite. A few months later, NASA (National Aeronautics and Space Administration) was born.

# The 1960s

The 1960s are known for great discoveries in astronautics. The Russians put a man in space in April of 1961. The Russians called this man a **cosmonaut** (sailor of the heavens). A month later, the U.S. sent Alan Shepard into space, and Americans called him an **astronaut** (sailor of the stars).

When Alan Shepard got back to Earth, the president of the United States, John F. Kennedy, decided that the U.S. should try to put a man on the Moon. No one laughed at him the way they had laughed at Goddard 40 years before. Eight years after President Kennedy publicly stated his desire to put a man on the Moon, an Apollo spacecraft sent men to the Moon. The United States had finally done something first! They were very excited. When Neil Armstrong stepped on the Moon he said, "That's one small step for man, one giant leap for mankind."

The United States beat the Russians to the Moon, but the Russians beat the Americans again a few years later by building the first space station. The Russian space station was called **Salyut**. What is a space station? It's a giant satellite that people can live in. The living area on the first space station was smaller than many people's closets. Many of these small space stations were built and sent into space by the Russians and the Americans.

A photograph of Neil Armstrong on the Moon.

Many years later, the Russian government fell apart, and the U.S. was no longer in a race with them. Americans and Russians are now friends and help one another to learn about space and astronautics.

Space stations are a great way for scientists to study space. Space stations orbit around Earth so quickly that they can look at much of the universe in a few hours. In fact, night comes about every hour on a space station. That means space station astronauts get to see the side of Earth having night and the side of Earth having day in about 2 hours!

Experiment growing White Spruce trees on a space station.

The greatest and most important use for space stations is that science experiments can be done in space! I'm sure you're wondering why someone would want to do experiments up in a little space station rather than right here on Earth. The reason is that when experiments are done without the influence of gravity, the results are different. Gravity affects everything: how plants grow, how fire burns, how germs spread, how chemicals mix together. The experiments done on the space station help us better understand how things work in the absence of gravity. This helps scientists make better medicines, equipment, products, and even toys!

I bet you didn't know that NASA makes toys. NASA science has led to lots of toys. Because NASA understands how to make things fly well, a toy company hired them to build a toy glider that would go a long way before hitting the ground. NASA also developed the controllers that many children use to play video games. These controllers are just like the ones used to fly spaceships to the Moon!

Space stations are pressurized with oxygen, heat, air conditioning, electricity, and many of the comforts of Earth. Inside the space station, the astronauts or cosmonauts don't have to wear a spacesuit. They are protected from the harsh environment outside. It's just like a home away from home.

## The International Space Station

The best space station ever built was assembled in 1998. This space station is called The International Space Station. *International* means between two or more nations. NASA used the word *international* because many nations helped NASA build the International Space Station, including Russia.

A photograph of the International Space Station orbiting Earth.

The International Space Station is like a home and science laboratory all in one. The people living on this space station are scientists who are trained as astronauts. NASA is always looking for expert scientists to do experiments in space. They need people who know a lot about life science, earth science, space science, gravity science, engineering, and space product development (making clever things). Almost any area of science can be useful to NASA. Do you like to do experiments? Perhaps one day you will do your experiments in space.

The scientists living in the space station have to get used to all kinds of changes from life on Earth. It takes them only 90 minutes to orbit Earth, so the sun rises and sets every 45 minutes. With the sun rising and setting like that, deciding when it's bedtime is a little more difficult. God made us to be on a 24-hour schedule. We operate best on the schedule God created for us, so even the astronauts try to keep a 24-hour schedule every day.

What is it like to live with no gravity on the space station? Scientists can simulate, or imitate, almost everything you will feel in space. However, on Earth it is very difficult to simulate not being affected by gravity. You can do it for short times when riding in an airplane that is falling. While the airplane is falling, you feel no gravity. In fact, that's why the astronauts don't feel any gravity in the space station. It's not that Earth's

This astronaut is floating because both she and the spacecraft she is in are orbiting Earth. This means both she and the craft are falling, so she experiences no gravity, even though gravity holds the craft in orbit.

gravity doesn't exist there. It certainly does. That's what holds the space station in orbit. However, the astronauts don't *feel* the gravity because they are falling with the space station.

Suppose you were on an elevator, and the cable that holds the elevator suddenly breaks. Oh no! The elevator would start falling down the shaft, and you would be falling right along with it. Do you think that your feet would touch the floor of the elevator during the fall? They would not. You see, since you would be falling at the same speed as the elevator, you would never reach the elevator's floor. As a result, you would float in the elevator, as if there were no gravity. That's what happens on the space station. The space station is constantly falling as it travels around Earth. The astronauts are falling with it. Therefore, the astronauts float, experiencing no gravity.

No gravity is called **zero gravity**, or **0-gravity**. It is impossible to really prepare someone for experiencing zero gravity over a long time. Astronauts can experience zero gravity for a short time by getting in a plane and allowing the plane to fall to Earth for a while. However, they can only do that for a brief moment before pulling up to avoid crashing into Earth. To try and prepare the astronauts for longer exposure to zero gravity, NASA trainers put the astronauts in tanks of water. Since the water makes our bodies float, it is a bit like experiencing no gravity. However, it's not exactly the same because you can control your movement by pushing against the water. In zero gravity, there is nothing to push against to keep you steady or to help you move where you want to go. You simply float, like a balloon thrown into the air.

This astronaut is experiencing 30 seconds of zero gravity using a NASA craft that goes up into the air and then falls toward the ground. This craft is often called the **vomit comet** because falling like that can make people sick at first.

In space, an astronaut flies and tumbles around like Peter Pan. It's hard at first to get used to, but once the astronaut settles in, life in space seems normal. An astronaut gets used to not having any weight. When astronauts return to Earth, however, they suddenly feel like they weigh as much as 12 trucks. That's because they are not used to feeling gravity pull on them.

Gravity keeps everything in place and working as God designed it. Did you know that your blood flow was designed to use the force of gravity? On Earth, gravity keeps the blood flowing down to your legs. We need a lot of blood in our legs because they do so much work for us. In space, our legs aren't as important anymore. An astronaut hardly uses them at all. Astronauts must get lots of exercise to keep them from becoming too weak to support them when they come back home. Without gravity, the blood gathers into the chest and upper body, instead of flowing down into the legs and feet. That's why astronauts in space can get large upper bodies and

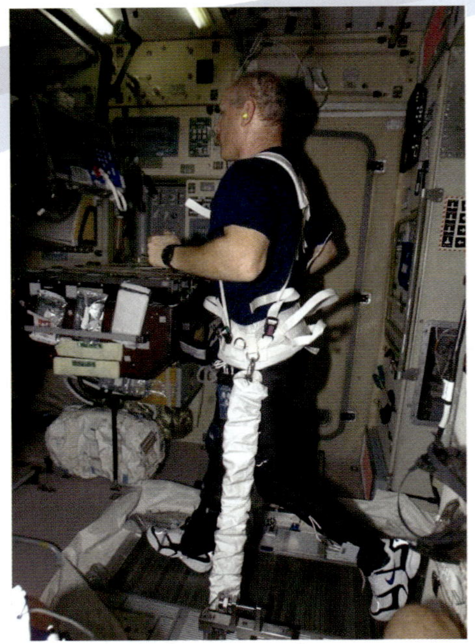
An astronaut exercising on the space station. Notice the straps that keep him from floating off the treadmill.

skinny legs. This is called **chicken-leg syndrome.** Lots of exercise helps to avoid this, so astronauts exercise each day they are in space.

Did you know that water and other liquids don't flow downward in space? In space, water floats into the air, like water-filled balloons that will break on whatever they hit. Just imagine if you spilled your milk while having lunch on the space station. Your milk would float about everywhere! Liquids ruin electrical equipment. Astronauts have to be very careful with liquids in space. They get drinks by sucking water out of a water bag that looks like a juice bag. They can also get water from a tube attached to the wall. If water droplets somehow get into the air, these water droplets must be caught or they could ruin the equipment. Special vacuum cleaners suck the water out of the air.

Liquid isn't the only potential problem in space. Astronauts can't set their food on a plate—it will simply float off! Food is kept in pouches. It is usually dried, with no crumbs. Imagine spilling crumbs in space. They would float in every direction. It would be difficult to clean up, and, like liquid, might get into the equipment and ruin it! The equipment on the space station is very delicate. Astronauts must take special care of the space station. They depend on everything to work perfectly so that they can stay alive and well. Crumbly food is not allowed.

An astronaut posing with his Thanksgiving meal aboard the International Space Station.

Fresh fruits and vegetables are also sent up to the astronauts in the space station, but it is very expensive to send shipments to space. It's not like sending a package to Grandma: the delivery man must be a trained astronaut, and millions of dollars are spent launching the delivery out of our atmosphere to the space station. Getting supplies of food, water or anything else is also very tricky because the space station is moving quickly in its orbit around Earth. The spacecraft making the delivery must be guided very carefully so that it can dock with the space station. As you might imagine, shipments aren't sent very often, so the space station carries a lot of supplies on hand.

Astronauts on the space station sleep in sleeping bags that are attached to the wall. They zip themselves in and strap themselves down so they will not float away in the middle of a dream.

These astronauts are getting ready to sleep. The beds don't look very comfortable, do they?

Using the restroom is also a lot different in space. Our toilets use gravity to work. Flushing the toilet only works because gravity pulls the water down. Since nothing on the space station is affected by gravity, there is no *down*. In space, the toilet is attached to the astronaut and gently sucks away all waste. That may sound awful to us, but to astronauts, it's just a normal part of life.

Since not knowing which way is up can bother the astronauts, space station builders, or engineers, place drawers, closets, equipment, and signs so that the astronauts know which way is *up*. Astronauts need to know this because they are often upside down, sideways, and doing tumbles in the air. An astronaut could just as easily walk along the ceiling as on the floor.

**Take a moment to share with someone what life in on the International Space Station would be like.**

# Building the International Space Station

Much of the International Space Station was put together up in space. The pieces, called **modules**, were built on Earth and assembled in space. Rockets and space shuttles, loaded with parts and pieces of the space station, were launched into space, so that the materials could be delivered to astronauts trained to build the space station.

Launching the heavy space station modules out of our atmosphere is a very difficult and expensive task. Once in orbit, however, the giant pieces of the space station become weightless! You can move a module piece, weighing as much as a building on Earth, like it weighs nothing at all.

A photo of NASA's Space Shuttle launch STS-88 delivering the Unity module to the Space Station.

Leaving the space station to work on the outside is called a **spacewalk** or an **EVA** (extravehicular activity). Space walks are very, very dangerous. When an astronaut leaves the inside of the station to work outside, he is exposed to extreme temperature changes. Without an atmosphere to keep the temperature stable, it is burning hot when the sun shines upon the space station, and it's terribly cold when the sun is not shining. The sun's solar winds are also quite dangerous. Do you remember the magnetosphere that shields Earth and keeps the solar winds directed toward the North and South Pole? Beyond the protection of the magnetosphere, the sun's harmful radiation coming from an unexpected solar wind could harm the astronaut very badly. Also, there is no protection from all the dust and rocks constantly hitting our atmosphere. Those little meteoroids continually rain on space-walking astronauts, like a constant sandstorm. Even a pebble-sized meteoroid can case severe injury, and some of the meteoroids are as big as a golf ball! Despite all these dangers, space walkers enjoy the excitement of being out of the ship.

A great deal of training is necessary to be a space station construction worker. Working on the International Space Station while in space must be done perfectly, or it could be dangerous for those living in the space station. This is especially hard work in a big bulky spacesuit, with no gravity to keep the astronaut stable. If the astronaut wants to turn, he must move his whole body. Also, the line that attaches the astronaut to the space station and keeps the astronaut from floating away can get in the way of the delicate work. The astronaut building the space station has a dangerous and important job.

# Becoming a NASA Astronaut

Being an astronaut is great fun. You might think you have to study space and astronomy in school to become an astronaut. Actually, NASA hires lots of people to become astronauts. Medical doctors, scientists who work in all sorts of different fields, and all kinds of engineers are employed by NASA. There are so many areas of science that are needed in space studies that many fields of science are good preparation for becoming an astronaut.

An astronaut waving at the camera while he is on a space walk outside of the International Space Station.

If you want to be an astronaut, it is important to take lots of math and science classes in school. Don't rush this part. Take your time and learn about Earth, its rocks, and its plants. Study living things, animals, and how bodies work. Study energy, motion, and electricity. All of these things are important fields of science, and NASA needs people trained in all of them. After you have studied many areas of science, decide which is your favorite. You will become an expert sooner than you realize! If your dream is to become an astronaut, you can do it if the Lord so wills with a lot of hard work, perseverance, and prayer.

A photo of an astronaut clowning around. He has just completed a spacewalk in which he successfully retrieved 2 broken satellites. They were loaded into the space shuttle to be brought back to Earth. I guess he was trying to make a little extra money, but there aren't many potential buyers up there.

## Seeing the International Space Station

Did you know that you can actually see the International Space Station in the night sky? It is close enough to Earth that it appears like a big star. However, it moves very quickly across the sky, orbiting Earth once every 90 minutes. Go to **www.apologia.com/bookextras** and you will find a link that takes you to a NASA program. This program will tell you when the International Space Station will be visible in your area and where to look for it.

# Activity 14.6

## Let's Visit the Planets!

To travel to the planets takes a long time, even in a very fast spacecraft. Let's do some math to figure out how old you would be if you travelled to each of the planets in our solar system. Suppose your spacecraft can travel at 10,000 miles per hour. That would be a very fast spacecraft, indeed. Follow the directions to find your age when you reach each planet. You will be surprised at how long you will be traveling!

**Older Students:** Do the following math exercise on your own to calculate how long it would take you to visit every planet in our solar system.

**Younger Students:** Parents, this is a fun activity to do with your child. Get out a calculator!

| | |
|---|---|
| The day you leave Earth is your next birthday. | How old are you the day you leave? |
| You arrive on Mercury in 8 months. | How old are you when you get to Mercury? (your age + 8 months) |
| You arrive on Venus 4 months later. | How old are you when you reach Venus? (your last age + 4 months) |
| You arrive back on Earth 4 months later and say hello to your family before you immediately set off for Mars. | How old are you when you reach Earth? (your last age + 4 months) |
| You arrive on Mars in 7 months. | How old are you when you reach Mars? |
| You arrive on Jupiter 3 years and 11 months later. | How old are you when you reach Jupiter? |
| You arrive on Saturn 4 years and 7 months later. | How old are you when you reach Saturn? |
| You arrive on Uranus 10 years and 2 months later. | How old are you when you reach Uranus? |
| You arrive on Neptune 11 years and 7 months later. | How old are you when you reach Neptune? |
| You arrive at Pluto 10 years later. | How old are you when you get to Pluto? |
| It will take 40 years and 10 months to make it back home. | How old will you be when you return to Earth? |

# Activity 14.7

## Build a Model Space Station

This project requires a lot of creativity, tape, glue, and stuff you can find around your house. You are going to build a model of the International Space Station.

You can use wires, paper towel rolls, empty soda bottles, craft sticks, straws, lids, and anything else you think might make a nice space station model. Try to make your model look as much like the International Space Station as possible. When you are finished, send Apologia Educational Ministries, Inc. a picture, and we will publish it on the course website! Be sure to include your name and age when you send us your picture.

## What Do You Remember?

Why do you see different stars during different times of the year? Which group of stars is always present in the night sky of the Northern Hemisphere? What is the name of the North Star? What is special about the star named Sirius? What is a black hole? What is a supernova? Describe the 3 star categories. What is a galaxy? In which galaxy is Earth? What is the shape of our galaxy? What is a constellation? How are constellations used today? What did Neil Armstrong say when he stepped on the Moon? What is a space station? How do you become an astronaut? What was your favorite part of this lesson?

We are now done studying God's magnificent universe together! Now, when scientists or the news discuss the sun, the stars, the planets, comets, asteroids and shooting stars, you'll know exactly what they are talking about. Just think about all that you've learned! You can now call yourself an amateur astronomer. Perhaps one day you'll take your studies even further, maybe even into space! Whether or not you decide to become an astronomer or an astronaut, you truly have grown so much in your knowledge and understanding of science. You should be proud of yourself! And always remember, in whatever you do, reach for the stars!

# APPENDIX

# SUPPLY LIST

## Lesson 1

- Adult supervision
- Balloons of many sizes and colors
- Scissors
- Thumbtacks or tape
- Thread, ribbon, or string
- Markers
- Measuring tape (If you do not have a measuring tape, you can use string cut to the lengths listed in the project.)
- Construction paper

## Lesson 2

- Adult supervision
- String
- A magnifying glass
- 1 inch pad of butter
- Paper plate
- A flashlight
- A globe (or round ball)
- Small ball (smaller than globe)
- A box
- Scissors
- White paper
- A pin or needle
- Tape
- Aluminum foil

## Lesson 3

- Adult supervision
- Small bowl
- Newspaper
- Flour
- Several pebbles of different sizes
- A marble
- A pencil
- 1 T flour
- 1 t salt
- 1 t water
- 1 drop blue food coloring
- 1 drop green food coloring
- 3 drops red food coloring

## Lesson 4

- Adult supervision
- 1 T butter or margarine
- 1 T flour
- A small plate
- A Thimble
- Small box
- Strong paper towel
- Tape
- Paper
- 1/4 C water
- 1/4 C flour
- Bamboo skewer (or a long, skinny stick with a point)
- Markers, crayons, or colored pencils
- Ruler (or measuring tape)

## Lesson 5

- Adult supervision
- Paper
- Flashlight
- Lamp
- Globe
- A cork
- Permanent marker
- A lid from a yogurt or sour cream container (with high lip.)
- A sewing needle
- A magnet

## Lesson 6

- Adult supervision
- Stick
- Lamp
- Lightly-colored ball (like a ping pong ball or white Styrofoam)
- Compact disc (CD)
- 2 magnifying glasses (One should be stronger than the other. Reading glasses will work also.)
- Construction paper
- Tape (Masking tape and duct tape work best.)
- Scissors
- Tape measure
- Paper with writing or an image on it

## Lesson 7

- Adult supervision
- Small bowl
- Rocks (optional)
- Alka-Seltzer® tablet
- Red and yellow food coloring
- ½ C flour
- 2 T of salt
- ½ tsp. cooking oil
- 2 T of water

## Lesson 8

- Adult supervision
- Paper
- Markers
- Tape
- Large open space

## Lesson 9

- Adult supervision
- 2 plastic bottles
- Water resistant tape (Electrical or duct tape works best.)
- 1-inch washer
- Water

## Lesson 10

- Adult supervision
- Alka-Seltzer® tablets
- Eye protection (such as safety goggles or glasses)
- Empty water bottle
- Water
- Tape
- Paper
- Scissors
- Clay plug
- Paper towel

## Lesson 11

- Adult supervision
- A glass jar
- A match
- Ice
- Large Ziploc® bag
- Hot water

## Lesson 12

- Adult supervision
- 2 T powdered sugar
- ½ C whipping cream (whole milk or half-and-half will work)
- ¼ t vanilla
- 6 T rock salt
- 1 pint-size Ziploc® plastic bag
- 1 gallon-size Ziploc® plastic bag
- Ice cubes

## Lesson 13

- Adult supervision
- 2 small balloons
- 10 inches of string
- 2 small rocks (1 inch in diameter)
- Water
- Dirt
- Eye dropper
- Pie plate
- Magnifying glass

## Lesson 14

- Adult supervision
- Dark-colored umbrella
- White chalk (or paint)
- Balloon
- Measuring device (such as a ruler or tape measure)
- Cardboard
- Clear plastic wrap
- Tape
- Scissors
- Pen (or marker)

# WHAT DO YOU REMEMBER? ANSWER KEY

Your child should not be expected to know the answer to every question. These questions are designed to help you review important concepts with your child. The questions are in plain type; the answers are in bold , but please keep in mind that answers could vary and still be correct. We recommend you spend the most time talking about the final question in each lesson.

## Lesson 1

Why did God create the stars and planets?
**As a calendar, to help the birds know when to fly south, to keep Earth stable, for His own glory.**

What are the names of the planets? **Mercury, Venus, Earth, Mars, Jupiter, Saturn, Uranus, Neptune.**

Do you remember the name of the astronomer who first said that Earth revolves around the sun? **Copernicus.**

What is the name of the astronomer who learned how to study space with a telescope? **Galileo.**

What is the name of America's space agency and what does it do? **National Aeronautics and Space Administration (NASA) builds equipment for studying space, trains astronauts, and sends spacecraft into space.**

What is the difference between a natural and an artificial satellite? **GOD made natural satellites like the moon. People created artificial satellites like spacecraft.**

What was your favorite part of this lesson? *Answers will vary.*

## Lesson 2

How many Earths would fit inside the sun? **1,297,126.** *Most children will answer, "Over 1 million," and this is acceptable.*

How many miles away is the sun? **Roughly 93 million miles.**

Can you explain the difference between revolving and rotating? **Revolve means to go all the way around, orbiting an object. Rotate means to spin, which turns night into day.**

Does the sun have any satellites? **Yes, everything orbiting the sun is a satellite.**

Explain what sunspots are. **Places on the sun that are cooler than the rest of the sun.**

Do you remember why you see color? **Different objects reflect different colors of light that bounce into your eyes.**

Can you explain what a solar eclipse is? **When the Moon gets in between the sun and the Earth, the Moon can block the sun's light, making it dark even during the day.**

What was your favorite part of this lesson? *Answers will vary.*

## Lesson 3

Is it hot or cold on Mercury? **It is BOTH hot and cold! During the day it is over 800 °F and at night it is almost -300 °F.**

What does the sky look like from the surface of Mercury? **Black.**

Why? **An atmosphere makes the sky have color. Mercury has no atmosphere.**

How long is a day (one rotation) on Mercury? **59 Earth days.**

How long is a year (one revolution around the sun)? **88 Earth days.**

Does Mercury orbit the sun in a circle or an oval? **An oval.**

What is the shape of Mercury's orbit called? **Elliptical.**

What kind of planet is Mercury: terrestrial or gaseous? **Terrestrial.**

What does the surface of Mercury look like? **Mercury has lots of craters.**

What are some reasons it might look like this? **Asteroids from outer space may have hit it.**

When is the best time of the year to see Mercury without a telescope? **Early in the morning or early at night**

Why? **It is close to the sun, so it tends to rise and set with the sun.**

What was your favorite part of this lesson? *Answers will vary.*

## Lesson 4

What would it feel like on Venus? **Burning hot.**

What is the atmosphere like on Venus? **It is thick with lots of poisonous clouds.**

What is special about the rotation of Venus? **It is opposite that of most planets.**

Why did astronomers think Venus was Earth's twin? **It is close to the Earth and is about the same size.**

Why does Venus go through phases? **Because of its orbit around the sun.**

Have many spacecraft have visited Venus? **Yes, 26.**

Since we can't see through the thick clouds covering Venus, how do we know what the planet's surface looks like? **Scientists use radar, which can get through the clouds.**

What was your favorite part of the lesson? *Answers will vary.*

## Lesson 5

What are the 7 things that make Earth the only planet that can support life?

**In any order:**

**1. Its distance from the sun**
**2. Its size (mass)**
**3. Its rotation**
**4. Its atmosphere**
**5. Its tilt**
**6. Its land**
**7. Its magnetosphere**

Try to explain why those things help us to live on Earth.

1. **If we were closer to the sun, the water would dry up. If we were further from the sun, the water would freeze.**
2. **The size (mass) of the Earth gives it perfect gravity.**
3. **If we rotated faster, we would have winds too strong to survive.**
4. **Our atmosphere gives us oxygen to breathe. It is required to grow plants. It keeps our temperatures steady. It protects us from meteorites.**
5. **If we were not tilted, we would not have seasons and could not grow crops in the summer in colder regions.**
6. **The layers of our Earth are designed perfectly for us to live on.**
7. **Our magnetosphere keeps solar winds from burning us with harmful radiation.**

Why do we have different seasons? **The tilt of the Earth causes us to have more or less direct sunlight, depending on where we are in our orbit around the sun.**

What are the 4 major layers of Earth? **Crust, mantle, outer core, inner core.**

What was your favorite part of this lesson? *Answers will vary.*

## Lesson 6

Can you explain why the Moon has phases? **As it circles the Earth, we are able to see different parts of the daytime side of the Moon.**

What is a lunar eclipse? **When the Earth gets right in between the sun and the Moon, casting its shadow on the Moon.**

What is the atmosphere like on the Moon? **There is no atmosphere.**

What is the color of the Moon's sky during its daytime? **Black.**

How does the Moon affect Earth's oceans? **It pulls on the ocean, causing the ocean's tides.**

Why is that helpful to Earth? **Tides cleanse the shore and keep the ocean from being stagnant.**

Why are there footprints likely visible on the Moon? **Because there is no wind or rain to wipe them away.**

What was your favorite part of this lesson? *Answers will vary.*

## Lesson 7

What is the name of the biggest volcano in our solar system? **Olympus Mons.**

What is the atmosphere like on Mars? **Cold and thin.**

Describe the surface of Mars. **It is red and has craters, volcanoes, and rocks.**

What do you remember about the moons of Mars? **They look like rocks, and they orbit closer and closer to Mars every day. They are named Phobos and Deimos.**

How long does it take Mars to revolve and rotate? **687 Earth days to revolve, 24 1/2 Earth hours to rotate.**

What is the weather like on Mars? **Cold.**

Why do some astronomers think Mars would be a good place to visit and, perhaps, live? **Scientists want to build a community there because Mars is a similar planet to Earth.**

What makes Mars look red? **Rusted iron in the dirt.**

What was your favorite part of this lesson? *Answers will vary.*

## Lesson 8

What is another name for a comet? **Dirty snowball.**

What does a comet leave behind as it orbits the sun? **Gas, dust and dirt.**

What happens when a comet's dust particles enter our atmosphere? **It lights on fire and looks like a shooting star.**

What do people call meteors? **Shooting stars and falling stars.**

What is a meteor called when it hits Earth? **Meteorite.**

Where have many meteorites been found? **Antarctica.**

From which planet did some of the meteorites come? **Mars.**

Where is the asteroid belt located? **Between Mars and Jupiter.**

What is the Exploded Planet Hypothesis? **All the asteroids in the Asteroid Belt were once a planet that exploded.**

Can you give some reasons that this might be a correct hypothesis? **Craters are found mostly on one side of many planets and moons; comets look like asteroids and could be pieces of this planet; Mars has the deepest craters and is the closest planet to the asteroid belt; many moons look like comets and asteroids.**

What was your favorite part of this lesson? *Answers will vary.*

## Lesson 9

How does Jupiter protect our planet? **Jupiter's gravity attracts comets which might otherwise hit Earth.**

Why is Jupiter like a little sun?**Jupiter produces its own heat like the sun, has a host of satellites, and is mostly hydrogen and helium.**

What is the Great Red Spot on Jupiter? **It is a giant storm that has been raging for three hundred years and is bigger than the whole Earth.**

Why does Jupiter have stripes? **Jupiter's stripes are bands of clouds.**

Name Jupiter's 4 largest moons. **Ganymede, Europa, Callisto, and Io.**

Why are they called Galilean moons? **Galileo discovered them.**

Can you describe Amalthea? **Amalthea is a big jumble of rocks held together by gravity.**

What do you remember about the spacecraft, Galileo? **Galileo explored Jupiter, sending back information. It took 6 years to get to Jupiter and took a roundabout route. It also had a problem with its antenna.**

What was your favorite part of this lesson? *Answers will vary.*

## Lesson 10

What is Saturn made of? **Mostly hydrogen and helium.**

Why would Saturn be an unpleasant place to visit? **Saturn is freezing, has no solid surface, and has terrible storms and hurricane winds moving across the planet.**

Which planet is considered Saturn's twin? **Jupiter.**

What are Saturn's rings made of? **They are made of ice, many rocks, boulders, and a few shepherd moons.**

What do shepherd moons do? **They keep the rings in place.**

How many years does it take Saturn to orbit the sun? **It takes Saturn 30 Earth years to orbit the sun.**

Why does Saturn look as if it is being squeezed? **It is spinning so fast that the planet flattens at the poles and bulges at the center.**

What is the name of the spacecraft mission which is currently studying Saturn? **Cassini is the mission currently at Saturn to explore its moons.**

What was your favorite part of this lesson? *Answers will vary.*

## Lesson 11

What chemical makes Uranus blue-green in appearance? **Methane.**

Why was it so exciting to discover Uranus? **A planet had not been discovered since ancient times.**

Who discovered Uranus? **William and Caroline Herschel.**

How were they educated? **They were homeschooled**.

How long does it take Uranus to orbit the sun? **84 Earth years.**

How many moons does Uranus have? **27.**

What is Uranus's largest moon? **Titania**

The sun rises in the west and sets in the east on 2 planets; which 2 planets are they? **Venus and Uranus.**

What was your favorite part of this lesson? *Answers will vary.*

## Lesson 12

What chemical gives Neptune its blue color? **Methane.**

Why was Neptune discovered? **Astronomers were looking for it.**

What made astronomers think there was another planet beyond Uranus? **Uranus's orbit wobbled as if it was being pulled by another larger planet.**

How long does it take Neptune to revolve around the sun? **164 Earth years.**

What was the Great Dark Spot? **A storm on Neptune.**

How many moons does Neptune have? **13**

What are the names of Neptune's 3 biggest moons? **Triton, Proteus, and Nereid.**

What are geysers? **They are holes in the ground through which chemicals underground spew.**

Is water coming from the geysers on Triton? **Probably not.**

What spacecraft has visited Neptune? **Voyager 2.**

What was your favorite part of this lesson? *Answers will vary.*

## Lesson 13

What is the Kuiper belt? **A belt of icy objects orbiting the sun past Neptune.**

What are the differences between a planet and a dwarf planet? **Dwarf planets do not have enough gravity to clear their surrounding area of space clutter.**

Why do we use spacecraft to study the dwarf planets? **Because our telescopes cannot see the planets very clearly from Earth while a spacecraft can pass by the dwarf planet very closely, take pictures, and measure scientific data.**

What was your favorite part of this lesson? *Answers will vary.*

# Lesson 14

Why do you see different stars during different times of the year? **The Earth revolves around the sun and the night sky faces different sections of the universe in its orbit.**

Which group of stars is always present in the night sky of the Northern Hemisphere? **The Little Dipper.**

What is the name of the North Star? **Polaris.**

What is special about the star named Sirius? **It is the brightest star.**

What is a black hole? **A collapsed star that is so small and massive that its gravity pulls in everything (even light) near it.**

What is a supernova? **An exploding star.**

Describe the 3 star categories. **Temperature, brightness, and total energy output (size).**

What is a galaxy? **A large group of stars.**

In which galaxy is Earth? **The Milky Way.**

What is the shape of our galaxy? **Spiral.**

What is a constellation? **A pattern of stars in the sky that make a dot-to-dot picture.**

How are constellations used today? **They help us identify where stars, planets, and other celestial objects are located.**

What did Neil Armstrong say when he stepped on the Moon? **"That's one small step for man, one giant leap for mankind."**

What is a space station? **An artificial satellite where people live and work.**

How do you become an astronaut? **Take a lot of math and science classes. Specialize in your favorite area of science, because NASA needs scientists in every field.**

What was your favorite part of this lesson? *Answers will vary.*

# INDEX

## O



## P



## R



## S



## T



## U



## V



## W



## Z

# PHOTO CREDITS

**Photos by Jeannie Fulbright:**
Lesson 2: Activity 2.2-28, Activity 2.5-39, Lesson 4: Activity 4.2-63, Lesson 9: Activity 9.1 (both)-128, Activity 10.2-142 (bottom)

**Photos and illustrations from or using images from www.istockphoto.com:**
footer graphic and galaxy icon/activity kit icon-13-198, What Do You Remember (brain graphic)-24, 42, 52, 64, 82, 96, 110, 124, 133, 144, 154, 161, 174, 198, 203, Lesson1: planet collage -13, sundial -15 (top), migrating birds, 17 (bottom left), solar system-19, Lesson 2: The Sun title page-25, basketball and pepper-corn-26, magnifying glass-28, sun and earth-29, Earth rotating-31, rainbow-34, car-35, 36, solar eclipse-37 (top), solar eclipse -38(top & bottom left), annular eclipse-38 (bottom center), Lesson 3: title page-43, Lesson 4: title page-53, Lesson 5: title page-65, scales, baseball, tennis ball-67, tree, apple-68, globe-72, seasons-74, sledding-75, igloo-75, Earth's layers-76, magnet-77, Lesson 6: title page-83, moon-84, tides-91, football field-93, Lesson 7: title page-97, Lesson 8: title page-111, infographic elements-119, inner/outer planets-120, asteroid-122, Lesson 9: title page-125, lightning-127, Lesson 10: title page-135, Saturn-136 (top), Lesson 11: title page-145, Uranus-146, now showing sign-148, telescope-149, Lesson 12: title page-155, ice cream scoop-157, Neptune cartoon -161, Lesson 13: title page-163, planets-164, Lesson 14: title page, umbrella-178 (top), U.S. map-188 (bottom), starry night sky-189, Goddard-190 (top), Sputnik-190 (bottom), space badge-197

**Photos and illustrations from www.dreamstime.com:**
Moon Phases-16 (middle), Lesson 14: Big Dipper-177, star map-188

**Photos and illustrations courtesy NASA/JPL/Caltech/Goddard/SDO/ESO:**
Cover image, Lesson1: planet collage background-13, stars and planets-15 (bottom) solar system art -18, astronaut-20, 21 (all), Lesson 2: solar system-30, solar flare-32, sunspot-32, sunspot close-up-33, blue atmosphere-35, Moon's craters-38, Bailey's Beads-38 (bottom right), SOHO -40 (bottom), sun and Mer-cury-44 (all), Earth and Mercury-46, Mercury's surface-46, Mercury transit-48, Mariner 10-49, Messenger poster-50, MPO-51, Mercury craters-51 (both), Lesson 4: Venus-54, volcanos-54, Venus clouds-56, Venus lightning-56, Earth Venus twins-58, Mariner 2-60, volcano-61, radar images-61, Lesson 5: Earth-66, Earth in front of sun-66, astronaut-69, Earth atmosphere-70, meteorite-70, magnetosphere-77, AIM-80, Pale Blue Dot-81, Earth and moon-85, Lesson 6: astronaut-90, maria-90, comman module-95, ascent/descent-95, lunar rover-95, footprint-95, LRO-96, Mars-98, Olympus Mons-98, crater-99, Lesson 7: Mars sky col-or-100, Mars sunset-101 (with Texas A&M), meteorite-101, crater-101 (ESO/DLR/FU Berlin), solar wind earth-102, Mars moons-103, Martian orbit-103, Mars North Pole-104, Mars surface-105, Moving to Mars-105, *Red Planet* surface-107, rocks-107, Sojourner-108, Opportunity-108, spacecraft rovers-109, Human tracks-109, Lesson 8: Comet McNaught-113, Swassmann-114 (top right), Comet Borrelly-114 (middle left), comet passing sun-116 (top), Halley's Comet-116, Jupiter comet-117, meteorite-117 (bottom), me-teor International Space Station-118 (top), Kleopatra meteroid-118 (bottom), Hellas Basin-121, Lesson 9:

Jupiter-126, Great Red Spot-127 (bottom), Great Red Spot close-up-128 (top), Jupiter's rings-129, Jupiter's moons-129 (bottom left), Europa-129 (bottom right), Io-130 (top), Callisto-130 (middle), Ganymede-130 (bottom), Amalthea-131 (top), Galileo-131 (bottom), launch-132 (left), Galileo probe-132 (right), Lesson 10: Saturn's tilt-136 (bottom), Saturn and moons-137 (top), icy amonia-137, Saturn's colorized rings-138, Saturn's rings-139 (bottom left), close up of rings-139 (top right), moons and rings-139 (bottom right), Cassini mission artwork-140 (bottom), Saturn's moons-141 (top right), Titan-141 (middle left), Titan art-141 (bottom left), Titan's surface-141 (bottom right), Ligeia Marie-142, Centaur launch-143 (bottom left), Cassini launch-143 (bottom right), Lesson 11: Uranus rings -146 (bottom left), Mimas-147, Uranus moons-151 (top), Titania-151 (middle left), Oberon-151 (middle right), Miranda's surface-151 (bottom), Ariel-152 (top), Umbriel-152 (bottom), Lesson 12: Neptune-156 (top), Neptune's rings-156 (bottom), Great Dark Spot close-up-159, Triton and Neptune-160, Proteus-161 (top left), Nereid-161 (top right), QB1-165 (top), diagram-165 (middle), New Horizons-165 (bottom), objects-166, Ceres-168 (top), Dawn-168 (bottom), Pluto-169 (top), size comparison-169 (bottom), Pluto orbit-170 (top), Pluto's moons-170, Eris-171 (top), Makemake-171 (bottom), Haumea-172 (top), solar system-172 (bottom), Lesson 14: Deneb-176 (Gendler), black hole (Drawing by A. Hobart)-179, Crab Nebula-180, binary star system-181, SpiralGalaxy (NASA, ESA, and the Hubble Heritage)-186 (all), Armstrong on moon-191 (top), experiment-191 (bottom), International Space Station-192, astronaut floating-193 (both), astronauts-194 (all), launch-195, astronaut waving-196 (top), astronaut with sign-196 (bottom), International Space Station-198

**Illustrations using NASA images:**
Lesson 1: Copernicus models-20, Lesson 2: 25, Earth revolving-30 (left), Lesson 2: sun (in car illustrations) 35, 36, Lesson 3: Mercury orbit-45, Lesson 4: title page-53, Venus rotation-57, Venus phases-59, Lesson 5: Earth's rotation-69, Earth seasons-73, Earth on flier-81, Lesson 6: Phases of Moon-86, lunar eclipse-89 (top) Saturn V-94, Lesson 7: title page-97, Earth, moon, Mars comparison-100, solar wind mars-102, Lesson 8: comet, 112, comets' orbits-114 (bottom), Jupiter's rotation-129, Saturn's rotation-140, Uranus rotation-150, Lesson 12: Neptune orbit-158, Lesson 13: dwarf planets size comparison-173, orbit-178 (bottom)

**Photos published under the Creative Commons Attribution-ShareAlike 2.0 Unported:**
Lesson 1: Heavens_Ian Norman-http'/www.lonelyspeck.com-14, Lesson 5: Aurora Over Bute by Mark-78 (bottom), Lesson 6: Super Red Moon by halfrain-88

**Photos published under the Creative Commons Attribution-ShareAlike 2.0 Generic:**
Lesson 3: Venus, moon and Mercury after sunset, Raymond Shobe-49, Lesson 8: Comet Hale-Bopp, Philipp Salzgeber-112, Leonid meteor shower, Ed Sweeney-117 (middle), Lesson 14: MilkyWay by Forest Wander, www.ForestWander.com-185, Cassiopeia by Mike Durkin-187, Canis Major by RichardAshley -187

**Photos published under the Creative Commons Attribution-ShareAlike 3.0 Unported, 2.5 Generic ; (http:creativecommons.org/licenses/by-sa/3.0/):**
Lesson1: astrolabe, Poulpy-17 (bottom right), Lesson 7: biosphere, Shimada -106, Lesson 8: Hale-Bopp, Schnobby-116, Lesson 14: Orion Head to Toe by Rogelio Bernal Andrea-183

**Photos published under the Creative Commons Attribution 2.0 Generic (CC BY 2.0) (https://creativecommons.org/licenses/by/2.0/):**
Lesson 1: Stonehenge, LocoSteve-16 (bottom), Lesson 2: Mercury v. Sunspot-48 (top),
Lesson 5: Aurora, Carsten Frenzi-78 (upper left), Aurora, Greg Clarke-78 (middle left), Lesson 6: lunar eclipse by makelessnoise-89, waves,"Mixmaster" by "belindah-Thank You!-400000 Views"-92, Lesson 7: Arizona Crater by Graeme Maclean-101 Lesson 11: Like Summer Clouds_Håkan Dahlström-153

**Photos published under the Creative Commons Attribution 3.0 (CC BY 3.0) (https://creativecommons.org/licenses/by/3.0/us/):**
Lesson 1: NightSky, ESO astronomer Yuri Beletsky-15

**Photos from public domain:**
Lesson 1: Zodiac Mosaic, 17 (top), Lesson 2: Partial Eclipse, Matt Hecht-37 (bottom right), Aurora Borealis, US Air Force-78 (upper right), , Lesson 5: Aurora, Chris Denals, National Science Foundation-78 (middle right), Lesson 8: Meteor Crater, U.S. National Geographical Survey-118, Lesson 11: Johann Elbert Bode-152

**Photos used with permission:**
Lesson 4: Sunset, Mark Whitney-58, Lesson 10: Saturn's rings-Dr. David Heatley, 139